VERTEBRATE DEVELOPMENT

VERTEBRATE DEVELOPMENT

by

Harold W. Manner
Loyola University of Chicago
Chicago, Illinois

KENDALL/HUNT PUBLISHING COMPANY
DUBUQUE, IOWA

QL
959
.M27

Copyright © 1975 by Harold W. Manner

Library of Congress Catalog Card Number: 75-9182

ISBN 0—8403—1150—8

All rights reserved. No part of this publication may be reproduced, stored in a retrieval system, or transmitted, in any form or by any means, electronic, mechanical, photocopying, recording, or otherwise, without the prior written permission of the copyright owner.

Printed in the United States of America

Contents

Foreword, ix

Preface, xi

Part I. General Embryology

Chapter 1. Introduction, 2

What Is Embryology? 3 Types of Embryology 3 The Animals to Be Studied 5

Chapter 2. Gametogenesis, 6

Formation of Gametes 6 Formation of Sperm 6 Formation of Eggs 8 Morphology of the Sperm 10 Egg Morphology 12 Meiosis 13 Chromosomal Aberrations 17 Significance of Meiosis 18

Chapter 3. Fertilization, 20

Internal and External Fertilization 22 Importance of Fertilization 23 Chemistry of Fertilization 26 Development Subsequent to Fertilization 27

Chapter 4. Cleavage, 29

Holoblastic Equal Cleavage 33 Holoblastic Unequal Cleavage 34 Meroblastic or Discoidal Cleavage 35 Significance of Cleavage 36

Chapter 5. Blastula and Gastrula Formation, 39

Blastulation in an Isolecithal Egg 39 Blastulation in a Moderately Telolecithal Egg 39 Blastulation in a Heavily Telolecithal Egg 40 Gastrulation 43 Gastrulation in an Isolecithal Egg 44 Gastrulation in a Moderately Telolecithal Egg 45 Gastrulation in a Heavily Telolecithal Egg 48

Chapter 6. Neurulation, 52

Mesoderm and Notochord Formation in an Isolecithal Egg 52 Neurulation in an Isolecithal Egg 55 Mesoderm and Notochord Formation in a Moderately Telolecithal Egg 58 Neurulation in a Moderately Telolecithal Egg 59 Mesoderm and Notochord Formation in a Heavily Telolecithal Egg 60 Neurulation in a Heavily Telolecithal Egg 61 Student Review 62

Part II. Specific Embryology

Chapter 7. The Development of Amphioxus, 68

Egg Morphology 68 Fertilization 69 Precleavage Egg Symmetry 70 Cleavage Pattern 72 Blastula Formation and Egg Orientation 74 Gastrulation 76 Ectodermal Development 77 Mesodermal Development 80 Development of the Endoderm 82

Chapter 8. The Embryology of Teleosts, 85

Egg Morphology 86 Fertilization 87 Cleavage and Blastula Formation 88 Gastrulation 90 Fate Map of the Early Gastrula 90 Gastrulation Movements 91 Neurulation 94 Formation of the Primitive Gut 95 Early Mesodermal Differentiation 96 Formation of the Body Shape 97 Early Development of the Neural Tube 99 Early Development of the Primitive Gut 99 Early Circulatory Pattern 101 Later Development of the Teleost 102

Chapter 9. The Embryology of the Frog, 103

Morphology of the Egg 103 Fertilization of the Egg 104 Cleavage in the Amphibian Egg 106 Fate Map of the Amphibian Blastula 107 Gastrulation 108 External Development to 48 Hours 110 Internal Changes Within 48 Hours 114 Ectodermal Development 114 Endodermal Development 118 Mesodermal Development 120 Later Development 122 External Development 122 Posthatching External Development 124 External Changes at Metamorphosis 127 Posthatching Internal Development 129 Ectoderm 129 Development of the Brain 129 The Spinal Cord 131 The Peripheral Nervous System 133 Spinal Nerves 133 The Cranial Nerves 134 The Autonomic Nervous System 136 The Eye 136 The Ear 138 The Nasal Chambers 139 Other Ectodermal Derivatives 139 The Endodermal Derivatives 140 The Respiratory System 141 Other Foregut Derivatives 141 Midgut and Hindgut 142 Mesoderm Development 142 Somites 142 The Mesomere 144 The Pronephric Kidney 144 The Mesonephric Kidney 145 The Hypomere 146 Heart Development 146 Development of the Vascular System 148 The Arteries 148 The Cardinal System 150 The Development of the Posterior Vena Cava 150 Hepatic Portal System 152 Reproductive System 154

Chapter 10. The Embryology of the Reptile, 156

Egg Morphology 157 Fertilization 157 Cleavage 158 Formation of the Periblast 159 Blastulation 159 Gastrulation 162 Neurulation 163 The Formation of Body Shape 164 Later Development of the Turtle 167

Chapter 11. The Embryology of the Chick, 169

Egg Morphology 169 Fertilization 170 Cleavage 170 Gastrulation 171 Neurulation 176 The 33-Hour Chick 178 Nervous System 178 The 48-Hour Chick 180 Ectodermal Development 183 Gill Arch Development 185 Endodermal Development 186 Mesodermal Development 187 Other Circulatory Vessels 190 The Cleidoic Egg 191 The 72-Hour Chick 194 Flexures and Torsion 194 The Nervous System 196 The Brain 196 The Spinal Cord 197 Peripheral Nervous System 197 Sense Organs 198 The Otic Vesicle 198 The Nasal Pits 198 The Eye 198 Endodermal Development 199 The Oral Plate 199 Pharyngeal Changes 199 The 96-Hour Chick 201 Somite Differentiation 202 Foregut Derivatives 202 Development of the Heart 203 The Fate of the Aortic Arches 204 Blood Vessel Development 205 Five Days to Term 207 The Respiratory Tract 210 The Digestive Tract 210 The Circulatory System 211 The Excretory System 213 The Reproductive System 214

Chapter 12. The Embryology of the Mammal, 216

The Mammalian Egg 216 Fertilization 217 Cleavage 217 Blastulation 218 Gastrulation and the Formation of the Primary Germ Layers 219 Formation of the Endoderm 219 Formation of Mesoderm, Notochord, and Ectoderm 220 Postgastrular Mesodermal Development 222 Postgastrular Ectodermal Development 222 Postgastrular Endodermal Development 223 Formation of Extraembryonic Membranes 223 Implantation 226 Later Development of the Mammal 227

Chapter 13. In Retrospect, 229

Index, 231

Foreword

Harold W. Manner is no novice in the field of writing textual material for college courses in biology. Beginning with his *Elements of Anatomy and Physiology* in 1962, he has singly or with colleagues produced an impressive volume of carefully composed material including *Elements of Comparative Vertebrate Embryology* in 1964, *General Biology* in 1966, and finally *Laboratory Manual of General Biology* in 1966.

He is recognized widely among his peers as a capable scientist, researcher, writer, and professor. Another book bearing his name is not particularly unexpected. As he observes in his Preface, however, there should be something different enough in a new product to warrant its being. Much of what is different in this present book has to do with the author himself.

Dr. Manner is one of the most sought-after professors at Loyola University. Possibly he teaches more students than any of his colleagues. His own course in Embryology may enroll more than five hundred students. Beyond his university classroom and laboratory, he is also sought after as a speaker for youth gatherings and conclaves as well as before business and professional conferences and conventions. As these groups come together, many of them are oriented toward value-centered education, and are seeking a meaningful application to the problems and concerns of life of the truths ranging from those of the scientific laboratory to those of the theologian and moral philosopher.

It was not always thus. In 1968 the orientation of his own life was quite significantly changed. While attending the meetings of one of the leading scientific societies being held in Washington, D.C., he began a serious study of the Sermon on the Mount from Matthew's account in the Holy Bible. This led to a deeper analysis of the fuller content of the Scriptures, not only as they bear on man's relationship to God, but as they bear on the facts of observable science.

While Dr. Manner has experienced a deeply satisfying, personal rebirth in his life, he has not diminished his orderly, scholarly study, teaching, and writing in the biological sciences. His present work is ample evidence of this.

Unerring in his commitment to factual, well-documented content and conceptual arrangement, he has, nonetheless, presented the data so that the student has a viable option in choosing a model for explanation and understanding of origin and development of life forms. No student of true science can fault this conceptual framework. For many, it will be a new and welcome approach.

William Randolph Davenport
Campbellsville College
Campbellsville, Kentucky
March, 1975

Preface

There are many good embryology textbooks on the market today. It would be unwise to add another at this time, unless it were sufficiently different to warrant its existence. Science today is progressing rapidly. In educating our students we must keep abreast with the latest developments. In many embryology courses it has been the practice either to study only one form, such as the chick or mammal, or to study a few forms which are considered to be representative of the entire vertebrate subphylum. In many vertebrate embryology courses, the animals utilized are the frog, the chick, and the foetal pig. Embryology on the undergraduate level is not, and should not be, the same type of course designed for medical and dental students. It should not only be a thorough investigation into the principles underlying all vertebrate development, but, in addition, should unify the vertebrates by noting the similarities and dissimilarities of as many vertebrate ontogenies as time allows. A knowledge of the embryology of the chick is relatively useless unless it is related to the development of other vertebrate types. Many competent investigators recently have been adding to our knowledge of the descriptive embryology of the vertebrates. Two vertebrate classes, the osteichthyes and the reptiles, whose embryology is often omitted in a basic embryology text, have been included in this book.

The first part of the text is concerned with those general principles of embryology that underlie all vertebrate development. The second half incorporates these principles into a descriptive survey of six chordate types. The omissions are numerous, but it was felt that by presenting only the basic tenets to the student, the teacher would be free to expand upon these in the classroom. The same is true for the illustrations. All the line drawings in this book are original and are designed to show, to their best advantage, specific embryological details. For this reason the number of labels has been kept to a minimum. If the instructor feels that the diagrams are inadequately labeled, additional ones of his own choosing can be easily added.

No book can possibly be the work of one person. I am indebted to many who have worked diligently with me throughout the preparation of the manuscript. The original manuscript was read with care by Dr. Nelson Spratt, Jr., an embryologist for whom I have the greatest respect. His numerous suggestions have been most helpful. The chapter on reptilian development could not have been written were it not for the generosity of another fine embryologist, Dr. Chester Yntema, who allowed me unlimited use of both his laboratory and his slides of reptilian embryos.

Much of the chapter on Teleost development was derived from the work of Casimira Dewese in my laboratory. This embryological study was made possible by a generous research grant from the Procter and Gamble Co. of Cincinnatti, Ohio. Drawings are an extremely important part of any textbook. Louis Rintrona and James Hardwick, two former students, have painstakingly worked to create illustrations to explain concepts that are sometimes difficult to put into words. Mrs. Josephine Johnson, my untiring secretary, typed the final manuscript and caught many errors which might otherwise have gone unnoticed. Neil Rowe and the entire staff of the Kendall/Hunt Publishing Company worked very closely with me throughout the preparation of this book. I would certainly be remiss if I did not acknowledge the stimulation and encouragement offered to me by my students.

Above all, I acknowledge the help given to me by a loving Father, whose answer to the prayers of thousands of my brothers and sisters in Christ, made this book possible.

Harold W. Manner
Chicago, Illinois

PART I

GENERAL EMBRYOLOGY

Embryology is a vast field full of the intricacies associated with any sophisticated science. It cannot be comprehended if approached in its entirety. If a solid foundation of basic facts can be mastered first, it becomes much easier to comprehend the entire field. Although each embryo develops along specific lines, each with its own peculiar ramifications, certain processes of each are similar to many others. The first part of this book, therefore, is generalized. No specific embryo will be studied. Instead, embryological terms and processes will be explained. Each of these plays an important role later in this book, when the embryological development of specific vertebrates will be studied.

"A wise man will hear, and will increase learning; and a man of understanding shall attain unto wise counsels." Proverbs 1:5.

Chapter **1**

Introduction

Observing an embryo change from an apparently homogeneous sphere into an organized individual with a pulsating heart, a differentiated brain, and other equally developed organs, within the relatively short span of a few hours, cannot help but leave an indelible impression upon any serious student. The factors involved in this awe-inspiring transition are, for the most part, still little understood. It is possible, however, to describe with accuracy the sequences that occur throughout the developmental period. A better understanding of the vertebrates is obtained when the close similarities of their embryos are observed. Describing the embryology of selected vertebrates is the primary aim of this book. Perhaps the knowledge of what is happening within the developing embryo will stimulate some to join the rapidly growing ranks of experimental embryologists who are dedicated to the analysis of the factors involved in causing these developmental changes.

In the initial phase of the study of any science, the terminology appears overwhelming to many students. Of necessity there are many new terms which must be learned. A close examination of embryological terminology, however, reveals that a majority of the terms are meaningful combinations of various Latin and Greek roots. A knowledge of these roots facilitates the learning of the embryological terms.

It is difficult to study embryology from a textbook alone. For this reason the student should have at his disposal the embryos of as many different chordates as possible. Most biology laboratories are equipped with whole mounts and cross sections of frogs, chicks, and pigs. These should be studied in the laboratory as the various embryos are discussed. Only by comparing the actual embryo with the diagrams in this book will a full appreciation be gained of the events occurring during embryological development. No drawing can be more than an approximation. The drawings in this book are certainly generalized and can do no more than try to represent what is actually occurring in nature.

What Is Embryology?

It might be correct to consider the beginning of an animal's life to occur when a sperm from the male parent unites with an egg from the female parent. It might, however, be more accurate to consider the formation of these germ cells as a necessary prerequisite for fertilization, and hence an integral part of the animal's life. It is difficult, therefore, to determine precisely when the life of the individual begins. At the other end of the scale is death, the end of life. During the intervening time, each organism passes through a series of stages. After fertilization, a period of embryological development occurs. This terminates when the embryo hatches or is born. A period of larval development usually follows. At a particular period in life, the larval form is replaced by that of the adult. This replacement can be extensive, e.g., tadpole to frog, or less dramatic, such as the transformation from a child to an adult. The entire sequence of stages is referred to as the *ontogeny* of the individual (Fig. 1-1). Every part of this ontogeny can be studied from a developmental aspect. For the purpose of this book, however, *embryology* will be defined as the entire period beginning with gamete formation or *gametogenesis*, and ending with birth or hatching.

Types of Embryology

Now that we have defined the term "embryology," we must realize that there are different methods by which embryological development can be studied. Each is, in its own way, a very important aspect of the over-all science of embryology. One method is to observe what is occurring in a developing embryo, and where it is

4   Introduction

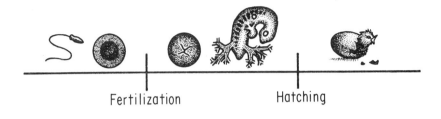

Figure 1-1. Ontogeny of the chicken, depicting the gamete formation, embryological development, early post-hatching, and adult stages of the life cycle.

taking place. We can watch, for example, the development of the brain, the arms and legs, and all the other organ systems of the body. We can also indicate the time of appearance of these various organs. Embryology so studied is called *descriptive embryology*. Basically, a descriptive embryologist is interested in describing what is happening, where it is happening, and at what time it is occurring in the developmental sequence. Other embryologists, however, are more concerned with analyzing the causes underlying this development. These men are not as much interested in the chronological sequence of events in the development of the limb as they are with such questions as: What is causing the limb to develop? Why does it grow? Where does it grow? What causes it to grow at a specific rate? This analysis of the causes of development forms the basis for *experimental embryology*.

Many different subsciences are included under the general heading of experimental embryology. Some experimental embryologists are concerned with the chemical nature of the causes underlying development. These men are *chemical embryologists*. Others believe that the developmental events occurring in the embryo are simply manifestations of the same basic physiological processes. This branch of experimental embryology is referred to as *developmental physiology* and includes the study not only of the basic embryological sequences but also of such developmental phenomena as regeneration and tumor growth in the adult animal.

Finally, certain scientists are not concerned primarily with either the descriptive or the experimental aspect of embryology. They are, however, interested in comparing the development of all vertebrates. This science of *comparative embryology* is valuable, for there are certain basic similarities underlying the development of all vertebrates. The embryological patterns found in the developing aortic arches, heart, respiratory systems, and kidneys are very similar in many embryos. This similarity aids the researcher by allowing him to use a more accessible animal form to study basic developmental problems.

The Animals to Be Studied

Each animal chosen for inclusion in this textbook are typical or representative of a particular vertebrate class. The *Amphioxus*, although not a vertebrate, does exhibit certain developmental characteristics which should be explored before passing on to the vertebrates. All eight classes of vertebrates will not be studied, for not all are available for study. The members of the Class *Placodermii*, for example, are all extinct. The four fish classes will be represented by the Fathead Minnow, *Pimephales promelas*, which currently is used extensively in water pollution research. This is an example of the Class *Osteichthyes*. The common grassfrog, *Rana pipiens*, will be studied as a representative of the Class *Amphibia*. This animal is particularly interesting because of its truly amphibious existence. The first part of its life is spent in the water, where it lives as an aquatic vertebrate. At *metamorphosis*, the transition from the larval to the adult state, the transition is from an aquatic to a terrestrial form. There are many examples of the Class *Reptilia*. The Painted Turtle, *Chrysemys picta*, was chosen because of the availability of its eggs. The egg of the reptile is interesting for it is constructed in such a way that it can be deposited in a nonwatery environment without drying out. This has enabled the reptiles to live in extremely arid places. The egg of the bird (Class *Aves*) is very similar to that of the reptile. Again, accessibility of eggs dictates the choice of the chicken as a representative of the class. In most embryological laboratories, the pig is studied as a representative of the Class *Mammalia*. It will be used as the type specimen in this book, although references will also be made to the human.

"Prove all things; hold fast that which is good." I Thessalonians 5:21.

Chapter 2

Gametogenesis

Formation of Gametes

All vertebrate development begins as the union of an egg and a sperm during the act of *fertilization*. This fertilization process has, therefore, as its primary requisites, an egg, produced by the female of the species, and a sperm, produced by the male. The process by which eggs and sperm are manufactured is called *gametogenesis*. The term itself means the origin of gametes, and gamete is a term that refers to both the egg and the sperm. When referring to the formation of sperm alone, the process is called *spermatogenesis*, whereas when the production of eggs, or ova, is primarily indicated, the term used is *oögenesis*.

Formation of Sperm

Spermatogenesis occurs in the testes of the male. The *testes* are organs located in most vertebrates in the anterior-dorsal part of the abdominal cavity. In mammals they are suspended in the scrotal sac. Figure 2-1 diagrammatically illustrates the process of spermatogenesis. Those cells that are destined to form sperm are called the *spermatogonia*. When these initially begin the process of spermatogenesis, they enlarge somewhat and are subsequently called *primary spermatocytes*. These soon divide to form *secondary spermatocytes*,

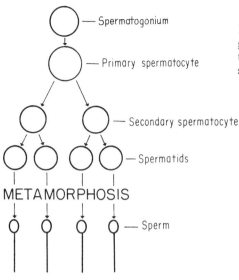

Figure 2-1. Diagram of spermatogenesis. Each spermatogonium in the male produces four functional sperm.

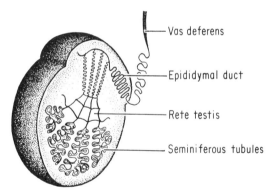

Figure 2-2. Section through a vertebrate testis. Sperm are manufactured in the seminiferous tubules. They then move through the rete testis to the epididymal duct, where they are stored. The vas deferens is the major duct conducting the sperm from the epididymis to the outside of the body.

which in turn divide again to form the four *spermatids*. The four spermatids undergo a change in their appearance known as *metamorphosis*. The result of this metamorphosis is the sperm proper. Note that each primordial germ cell gives rise to four sperm. Figure 2-2 is a diagram of a section through the testes. Figure 2-3 illustrates a cross

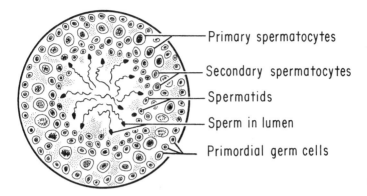

Figure 2-3. Cross section through a seminiferous tubule. As the process of spermatogenesis proceeds, the sperm move towards the lumen.

section through one of the *seminiferous tubules*. As the process of spermatogenesis proceeds, the developing sperm move toward the lumen of the seminiferous tubules. From here they enter the *rete testis* and are stored in the *epididymal ducts*. When it is time for mating, the sperm are carried through ducts, which vary in different animals, to the outside of the male body.

Formation of Eggs

The eggs are produced in a pair of glands called *ovaries*. These ovaries are located in the abdominal cavity of the female. A cross section through the ovary is illustrated in Figure 2-4. The cells that are destined to become eggs are referred to as the *oögonia*. When their development begins, they enlarge somewhat and are then called *primary oöcytes*. These divide unevenly, producing a *secondary oöcyte* and a *first polar body*. The *secondary oöcyte* further divides, forming an *oötid* and a *second polar body*. The first polar body also divides to form second polar bodies. The three polar bodies degenerate shortly after they are formed. The oötid then undergoes a period of metamorphosis resulting in a fully developed egg. Note that each oögonium in the female gives rise to one egg and three polar bodies. Compare this with the spermatogonium in the male which produces four functional sperm. This entire process is diagrammatically illustrated in Figure 2-5. As oögenesis proceeds, a layer of supporting

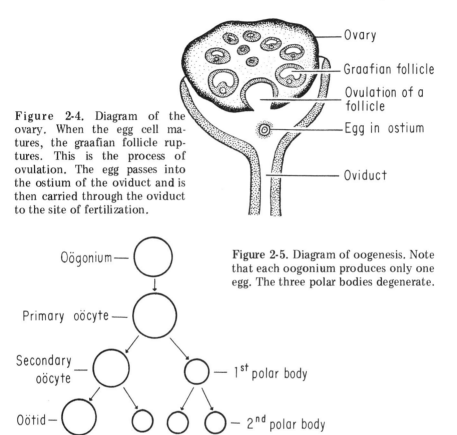

Figure 2-4. Diagram of the ovary. When the egg cell matures, the graafian follicle ruptures. This is the process of ovulation. The egg passes into the ostium of the oviduct and is then carried through the oviduct to the site of fertilization.

Figure 2-5. Diagram of oogenesis. Note that each oogonium produces only one egg. The three polar bodies degenerate.

cells surrounds the egg. These are called follicular cells, and the entire structure is known as a *graafian follicle.* This is illustrated in Figure 2-6. As development continues, the graafian follicle enlarges. The follicular cells secrete the female sex hormone, *estrogen.* As egg development nears completion, the graafian follicle reaches a point near the wall of the ovary. Here it ruptures, ejecting the egg to the outside of the ovary. This is the process of *ovulation.* The egg is actually liberated into the abdominal cavity. Here it is picked up by currents produced by the beating cilia of the ostium, the opening of the female ducts. The oviduct then conveys the egg, either to the outside, in those animals where fertilization is external, or to a site of fertilization in those animals where internal fertilization occurs.

Figure 2-6. Mature graafian follicle. The egg is supported by follicular cells.

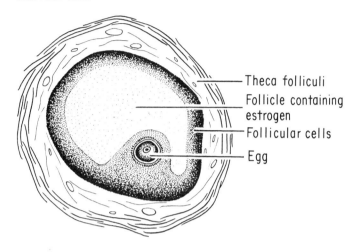

Morphology of the Sperm

The type of sperm produced by spermatogenesis varies from species to species. The size of the sperm may be as little as 0.020 mm in the crocodile or as large as 2 mm or more in Balanoglossus (a protochordate). The shape also varies, but in general it resembles a tadpole with an elongated tail. Sperm shape is, however, species specific. A few representative vertebrate sperm are depicted in Figure 2-7. The sperm can be divided into two distinct portions: a head anteriorly and a tail posteriorly. The head contains a darkly staining body, the *nucleus*, surrounded by a small amount of cytoplasm. At the anteriormost portion of the head is the *acrosome*, a dark body originating from the *Golgi apparatus* of the spermatid. Although the function of the acrosome is not completely understood, it has been postulated that it is concerned with the production of a secretion that is necessary for fertilization to occur. Figure 2-8 shows the head of the sperm. The tail of the sperm has three major divisions. For our purpose, we will refer to them as the middle, main, and end pieces. The entire tail is shown in Figure 2-9. The middle piece contains two centrosomes, a centrally located proximal centrosome, and a ring-shaped distal centrosome. Between these two centrosomes is an axial filament which arises from the proximal centrosome and passes through the ring-shaped distal centrosome. This axial filament is responsible for the property of mobility which all sperm possess.

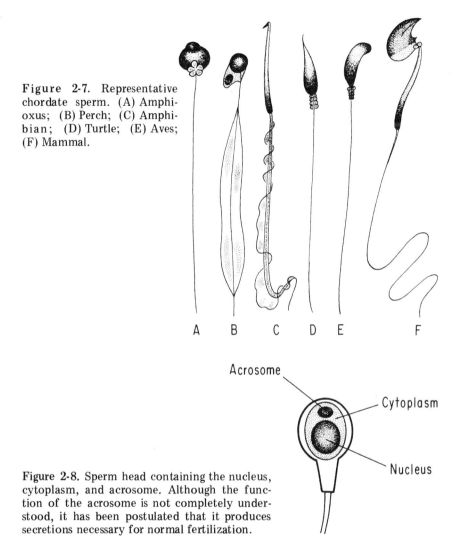

Figure 2-7. Representative chordate sperm. (A) Amphioxus; (B) Perch; (C) Amphibian; (D) Turtle; (E) Aves; (F) Mammal.

Figure 2-8. Sperm head containing the nucleus, cytoplasm, and acrosome. Although the function of the acrosome is not completely understood, it has been postulated that it produces secretions necessary for normal fertilization.

Surrounding the axial filament are numerous mitochondria or granules of mitochondrial origin. Surrounding the entire middle piece is a plasma membrane, the same plasma membrane that surrounds the head and other portions of the tail. The main piece is the longest portion of the sperm tail. It is composed basically of a long axial filament surrounded by the plasma membrane. The end piece is the naked axial filament.

12 Gametogenesis

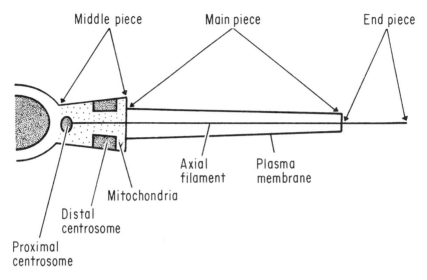

Figure 2-9. Details of sperm tail. The axial filament arises from the proximal centrosome.

Egg Morphology

The egg is the largest of the two gametes. Primarily, this is due to the fact that the egg must carry with it the food supply for the developing embryo. In many cases the entire period of development must be carried on without access to outside food. Even in those cases where extraembryonic food is supplied, such as in the mammal, a period of time does exist prior to the attachment of the embryo to its external food supply, during which the cells of the developing embryo require nourishment. The stored nutritive material in all eggs is called *yolk*. Vertebrate eggs differ both in the amount of stored yolk they contain and in the distribution of this yolk throughout the egg. Eggs can be classified on the basis of the amount of yolk they contain. The *ova* of the primitive chordates have very little yolk and are referred to as *meiolecithal* eggs. Those eggs with a moderate amount of yolk are *mesolecithal*. The mesolecithal type is characteristic of both the amphibians and the fishes. The *polylecithal* type of egg contains a large quantity of yolk. The reptiles and birds have polylecithal eggs. Mammalian eggs have little or no yolk and consequently are classified as *alecithal* eggs. In each case it must be remembered that this classification is based on the amount of yolk and not on the distribution of that yolk.

Eggs can also be classified on the basis of the *distribution* of the yolk. Three types exist (Fig. 2-10). The *isolecithal* type is an egg in which the yolk is distributed evenly throughout the egg cytoplasm. A *moderately telolecithal* egg is one in which the yolk is more concentrated at the vegetal than at the animal pole. An example of this type is the frog egg. Finally we have those types of eggs in which the yolk is extremely concentrated at the vegetal pole, so concentrated that only a small portion of cytoplasm exists at the animal pole of the egg. This disk of protoplasm is referred to as the *germinal disk*. The chick and the fish egg are both examples of *heavily telolecithal* eggs.

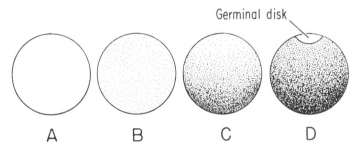

Figure 2-10. Types of vertebrate eggs. (A) Alecithal; (B) Isolecithal; (C) Moderately telolecithal; (D) Heavily telolecithal.

Meiosis

The normal sequence of meiosis is the same in both spermatogenesis and oögenesis. The classical pattern is illustrated in Figure 2-11. Using two pairs of chromosomes, a V-shaped and a rod-shaped pair, we note that in *prophase I* the *homologous* pairs attract one another and line up on the equatorial plate during *metaphase I*. The homologous pairs then split, giving four V-shaped chromosomes and four rod-shaped chromosomes. During *anaphase I*, two of the V-shaped chromosomes move toward one pole and two toward the opposite pole. The same movement occurs with the rod-shaped chromosomes. The cell then divides evenly, giving two daughter cells, each one having two V-shaped chromosomes and two rod-shaped chromosomes. Now if we follow the development of one of these, we will find that the V-shaped and rod-shaped chromosomes move toward the equatorial plate in *prophase II*. During *metaphase II* they are lined up on the equatorial plate. This is followed by *anaphase II*, during which each member of the pair of chromosomes begins to move toward the opposite end of the developing egg. During *telophase II* the cell again splits, and the result is either two oötids or

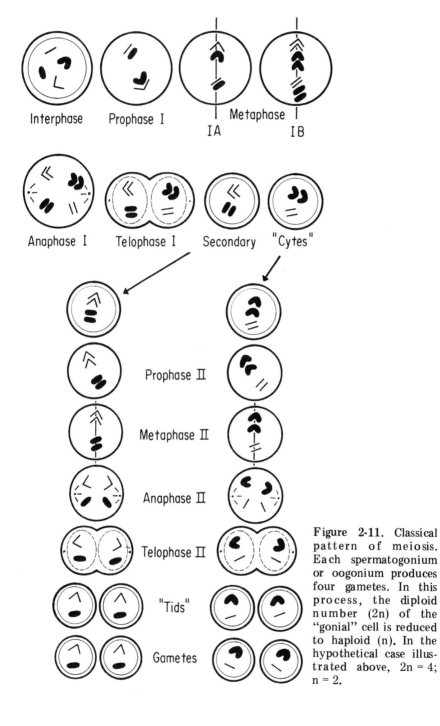

Figure 2-11. Classical pattern of meiosis. Each spermatogonium or oogonium produces four gametes. In this process, the diploid number (2n) of the "gonial" cell is reduced to haploid (n). In the hypothetical case illustrated above, 2n = 4; n = 2.

two spermatids having one rod-shaped and one V-shaped chromosome each. Note that in the classical pattern of meiosis represented here, the original diploid number of four is reduced to a haploid number of two.

This classical pattern would suffice for an elementary student in zoölogy, for it shows that the process of meiosis is, in reality, a reduction division, reducing the diploid number to haploid. In actual meiosis, however, the process is far more complicated than the classical pattern implies. Let us once again, therefore, follow the process of meiosis. The major difference between the actual and the classical pattern occurs in prophase I. Prophase I has a series of stages within it. To follow this let us once again take a cell having a diploid number of four. Prophase I is illustrated in Figure 2-12. During the initial part of prophase called *leptotene*, the chromosomes are long and threadlike. When they are in this condition, they are referred to as *chromonemata*. The number of chromosomes or chromonemata is equal to the diploid number or $2n$. The next stage

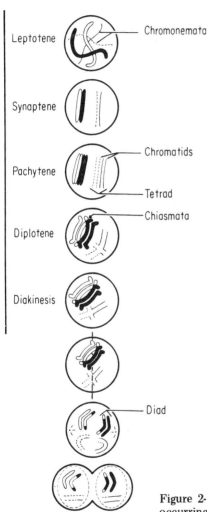

Figure 2-12. Prophase I substages. The changes occurring in prophase result in altered gene complexes. Consult the text for a description of each of these substages. The stages following prophase I are similar to those described in Figure 2-11.

16 *Gametogenesis*

is *synaptene* or *zygotene*. Here we have a synapsis or pairing of homologous chromonemata. The number of chromonemata still remains at $2n$. This stage is followed by *pachytene* during which the chromonemata shorten and become typical chromosomes. Each homologous chromosome splits, resulting in four *chromatids* or one *tetrad*. The term "tetrad" means four, and you will notice that for each original pair of chromosomes we now have four chromatids. It is at this stage that the phenomenon of crossing over occurs. To understand the process of crossing over, refer to Figure 2-13. The chromatids are extremely vulnerable to breakage. When breaks occur, there is the possibility that a recombination of chromosomal parts will be realized. The various types of breaks and recombinations are referred to as *chromosomal aberrations*.

Figure 2-13. Types of chromosomal aberrations. (A) Terminal translocations; (B) Intercalary translocations; (C) Terminal deletions; (D) Intercalary deletions; (E) Inversions.

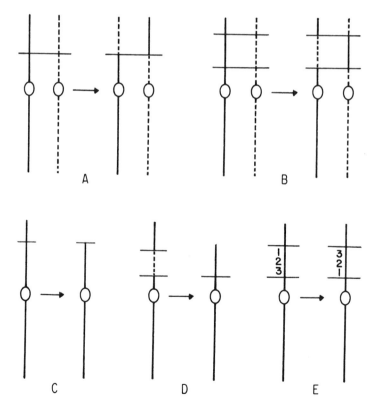

Chromosomal Aberrations

Many types of chromosomal aberrations occur during the prophase I stage of meiosis. The first of these is the *deletion*. In a deletion one or more breaks occur in a chromosome, and the breaking point is not healed. This break may occur at the end of a chromosome, deleting the chromosomal part lying near the end. Breaks also occur in the middle of the chromosome. When a deletion occurs at the end of a chromosome, it is referred to as a *terminal deletion* (Fig 2-13C). If it occurs in the middle of a chromosome, it is an *intercalary deletion* (Fig. 2-13D). Occasionally these deleted portions will return to a chromosome to heal in. In a few cases they heal inversely, with the posterior end now anterior and the anterior end now posterior. These are referred to as *inversions* (Fig. 2-13E). The third type, and the most numerous, are the *translocations*. In a translocation a part of one chromosome becomes firmly affixed to another chromosome. In other words chromosomal material transfers from one chromosome to another. If this occurs between two terminal pieces, it is known as a *terminal translocation* (Fig 2-13A). If it occurs between two pieces other than the end, it is an *intercalary translocation* (Fig. 2-13B). These are significant in future development in that they affect the position of the genes on the chromosomes. These genes, the particles of hereditary material that control the developmental sequence, are affected. This phenomenon is the *position effect* of the gene and in many cases can cause developmental change. At this point it may be germane to indicate that chromosomal aberrations occur spontaneously. They are not the result of *mutagenic agents*. We certainly have not exhausted the many possible types of chromosomal aberrations. For our purposes it is sufficient to indicate that these breaks do occur and that recombinations subsequently happen.

The next stage (Fig. 2-12), *diplotene*, possesses odd-shaped configurations brought about because of the chromosomal aberrations that occurred in pachytene. Even though parts of the chromatids are now on different chromosomes, they will try to synapse with their original mates. This results, in the case of the terminal translocation, in a configuration such as is found in Figure 2-14A, whereas the intercalary translocation appears as illustrated in Figure 2-14B. These points of crossing over are called *chiasmata*, and there are both a repulsion of the homologous chromosomes and a holding together at these chiasmata. In this condition the tetrads move to the metaphase plate during *diakinesis*. During diakinesis the chromosomes approach

18 Gametogenesis

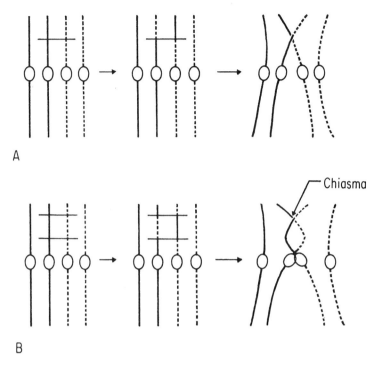

Figure 2-14. Types of tetrad configurations found on the equatorial plate in metaphase I. (A) Terminal translocation; (B) Intercalary translocation.

the equatorial plate and continue their period of shortening. Following metaphase I is anaphase I with these newly organized chromatids. Note now that there are only two chromatids in each configuration. They are, therefore, referred to as *diads*. Telophase I results in the splitting of the cytoplasm and the production of two secondary cytes, spermatocyte, or oöcyte. Each of these then goes through a similar cycle, going through prophase II, metaphase II, anaphase II, and telophase II as described under the classical pattern. The result is a spermatid or an oötid containing one of each of the original pair of chromosomes.

Significance of Meiosis

All the cells in the body are diploid. During normal mitosis each cell with its complement of chromosomes is duplicated. That means that every cell in the body has the same number of chromosomes. If this process of mitosis were to occur in the formation of gametes, the nucleus of both the egg and the sperm would be diploid. In the case of the hypothetical animal under consideration now, the egg would have four chromosomes and the sperm would have four chromosomes. At fertilization there would be a recombination of these, resulting in eight chromosomes. This of course would be a different number from the original parent. We know, however, that this does not occur. During fertilization a recombination occurs, and the resulting embryo is the same species as the parent. This can only mean that the egg and the sperm must have only half the parental number of chromosomes. It is for this reason that the normal mitotic process cannot occur during the production of gametes. The meiotic process is essentially a reduction division, reducing the original diploid number to half. This reduction by one-half is a very precise reduction in that one of every original pair of chromosomes occurs in the gamete, so that by recombining the gametes at fertilization, the original complement of chromosomes is restored in the embryo.

"And I gave my heart to seek and search out by wisdom all things that are done under heaven: this sore travail hath God given to the sons of man to be exercised therewith."
Ecclesiastes 1:13.

Chapter **3**

Fertilization

The act of fertilization occurs over a period of time, and any definition must take this into consideration. For our purposes, *fertilization* can be defined as the entire period beginning with the sperm's approach to the egg and ending with the fusion of the egg and sperm pronuclei. The nuclei of the gametes, having been produced by meiosis, are haploid, and to distinguish them from mature nuclei of other cells, they are referred to as *pronuclei*. The fertilization process is highly specific. Sperm will fertilize only eggs of the same species. The nature of this specificity is not well understood.

Both the egg and sperm have an extremely short life. The sperm, in most cases, will live only a few hours to a few days after they leave the male reproductive tract. The egg will also die within a period of two days. Even more startling is that, in many animals, the egg loses its ability to become fertilized in a short time. Neither the loss of fertilizability nor the short egg life can be attributed to a lack of food, for the supply of stored yolk is capable of sustaining the fertilized egg for a period of many days or even weeks in some cases.

In the process of fertilization, the sperm approaches the egg and its head penetrates the surface of the animal hemisphere. The tail drops off and plays no further role. This sperm entrance (Fig. 3-1) produces a change in the egg surface that prevents other sperm from entering. In a few vertebrates, e.g., the urodele amphibian, reptiles, and birds, this change does not occur as rapidly and many sperm enter. Only one, however, will fuse with the female pronucleus. The others degenerate. When only one sperm enters the egg, the process is called *monospermy*. When more than one enter, it is *polyspermy*.

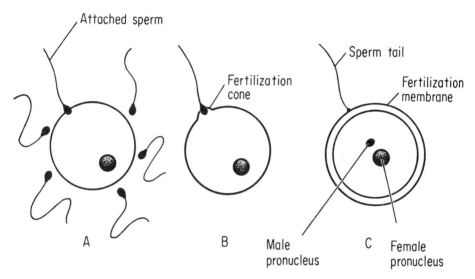

Figure 3-1. Sperm penetration. (A) One sperm has already attached itself to the egg surface and is seen in the process of penetrating the egg cortex. (B) The egg reacts by producing a fertilization cone which engulfs the sperm in a phagocytic manner. (C) Only the head of the sperm enters the egg cytoplasm. The tail remains outside. Once the sperm is inside, the vitelline membrane lifts from the surface of the egg, forming a fertilization membrane. The sperm and egg nuclei are called the male and female pronuclei respectively.

Much experimental work has been done on the factors involved in the process of fertilization. Many mysteries, however, still surround the act of fertilization. The sperm, for example, is not absolutely essential for fertilization and subsequent development to occur. The phenomenon by which an egg develops without a sperm is called *parthenogenesis*, and when this is artificially induced in the laboratory, it is referred to as *artificial parthenogenesis*. The method most commonly used to induce artifical parthenogenesis can best be illustrated in the most commonly used laboratory frog, *Rana pipiens.*

Ovulation in the frog is controlled by the pituitary gland. By injecting whole fresh pituitary glands into winter or hibernating

Figure 3-2. Artificial parthenogenesis in the frog. Eggs obtained from a hibernating or winter female frog by inducing ovulation with pituitary transplants are covered with a thin layer of frog blood (A). A needle is then used to prick the surface of each of these eggs, allowing blood plasma to enter the egg cytoplasm (B). If this is done correctly, approximately 5 percent of the eggs will begin cleaving within two hours.

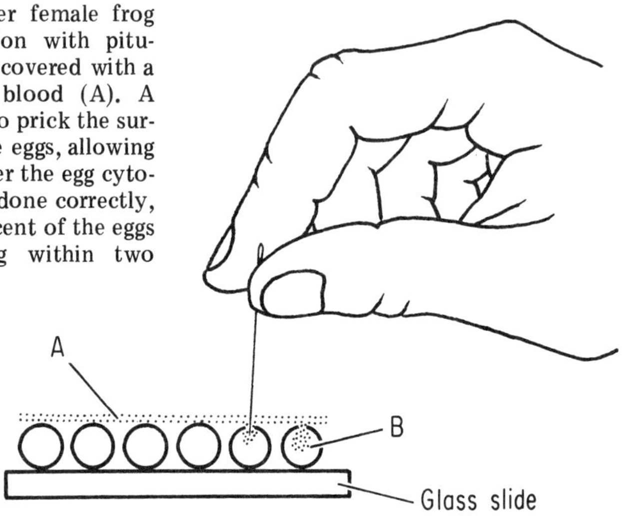

females, ovulation can be induced at almost any time of the year. The second prerequisite for successful artificial parthenogenesis is fresh frog blood. This is usually obtained by opening the abdominal cavity of a frog and cutting off the tip of the heart ventricle. This heart blood is then allowed to collect in the coelomic cavity. The ovulated eggs are then covered with this blood and immediately pricked with a needle, allowing some of the blood plasma to enter the cortex of the egg (Fig. 3-2). If this is done to a sufficient number of eggs, approximately 5 percent will be found in early cleavage stages within a period of two hours. Although this method is the one most commonly used, other agents have been utilized to induce parthenogenesis.

Internal and External Fertilization

Vertebrates can be classified on the basis of the place of development of the embryo into oviparous, ovoviviparous, and viviparous forms. The *oviparous* forms are those whose eggs develop outside the body. The *viviparous* forms are those in which development occurs within the body of the female. The young are born alive. In these vertebrates an exchange of nutrients and gases occurs between the maternal blood and the blood of the developing embryo through an intimately associated series of membranes called collectively the *placenta*. The *ovoviviparous* forms also have eggs that develop within the body of the female. However, there is no placental contact between the blood of the embryo and the blood of the mother. It is simply that the female is providing a safe place in which the eggs can develop.

Fertilization is either internal or external. In *internal fertilization* the actual contact between the sperm and the egg occurs somewhere within the reproductive tract of the female, usually in the upper third of the oviduct. In *external fertilization* the contact between the two gametes occurs outside the body. This latter type of fertilization is reserved for the water-dwelling forms, because the sperm, being motile cells, require some type of liquid medium in order to swim. External fertilization in aquatic forms is not, however, universal, for many water-dwelling vertebrates, such as sharks and some teleost fishes, exhibit internal fertilization.

The embryo of oviparous land animals must be protected from dessication. A different type of egg, the cleidoic egg (Fig. 3-3), accomplishes this by surrounding the embryo with an extraembryonic amniotic membrane which secretes a liquid that bathes the embryo during its developmental period.

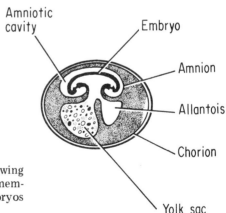

Figure 3-3. Typical cleidoic egg showing the four basic extraembryonic membranes which have enabled the embryos of land vertebrates to survive.

Importance of Fertilization

Although gametogenesis is certainly developmental, from a practical point of view, fertilization must be considered the starting point of embryonic development. This particular act has many significant facets which influence later embryonic progress. It restores the diploid number of chromosomes to the embryo, activates the egg to develop, and recombines the maternal and paternal genetic traits (Fig. 3-4). In addition, fertilization establishes the direction along which development will proceed. To understand this last statement, it is necessary to look more carefully at the events occurring during the fertilization process.

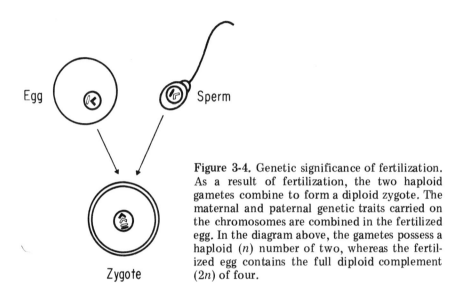

Figure 3-4. Genetic significance of fertilization. As a result of fertilization, the two haploid gametes combine to form a diploid zygote. The maternal and paternal genetic traits carried on the chromosomes are combined in the fertilized egg. In the diagram above, the gametes possess a haploid (n) number of two, whereas the fertilized egg contains the full diploid complement ($2n$) of four.

The egg undergoes two major physical changes shortly after sperm penetration. At the site of sperm contact a raised area of cytoplasm, the *fertilization cone*, is formed. This engulfs the sperm in a manner similar to phagocytosis (Fig. 3-5). After the sperm head and neck are inside the egg, the fertilization cone regresses. The *vitelline membrane*, which surrounds the egg, then lifts from the surface to become the *fertilization membrane*. This establishes a *perivitelline space* between the fertilization membrane and the egg (Fig. 3-6). This membrane, although selectively permeable to nutrients and gases, prevents other sperm from coming into contact with the egg.

The eggs of most vertebrates are in the metaphase stage of the second maturation division when sperm entrance occurs. While the

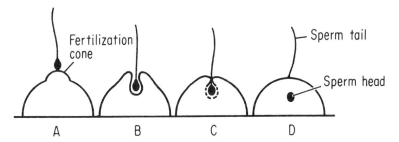

Figure 3-5. Formation of the fertilization cone. Upon contact of the sperm with the cortex of the egg, a raised area of cytoplasm called the fertilization cone develops (A). This fertilization cone engulfs the sperm head (B, C, D) in a manner similar to the process of phagocytosis in the amoeba.

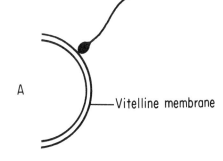

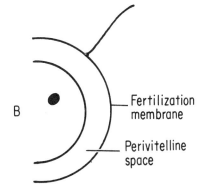

Figure 3-6. Formation of the fertilization membrane. Once the sperm head is inside the cytoplasm of the egg, the vitelline membrane, which is closely applied to the egg (A), lifts from the surface to become the fertilization membrane. This acts as a barrier to polyspermy. The space between the fertilization membrane and the egg surface is the perivitelline space.

egg is going through the last stages of oögenesis, the sperm begins to move toward the egg nucleus. This movement of the sperm appears to be "directed," and some investigators feel that it is due to a chemotaxic effect of chemicals liberated by the female pronucleus. During this movement toward the female pronucleus, the sperm may have to deviate from its *penetration path*. If it does, the new pathway taken is referred to as the *copulation path* (Fig. 3-7). The point of sperm entrance, penetration path, and copulation path have all been postulated as being responsible for establishing the primary plane of bilateral symmetry in the embryo. There is, however, no universal agreement on this subject for it is complicated by the knowledge that bilateral symmetry is established even in those eggs that develop without a sperm, i.e., by artificial parthenogenesis.

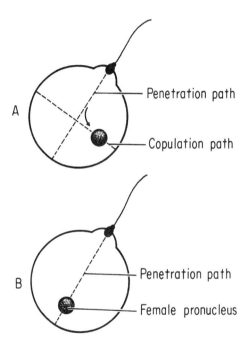

Figure 3-7. Sperm pathways during fertilization. In order to merge with the female pronucleus, the male pronucleus may have to deviate from its penetration path. This new path is the copulation path (A). If the female pronucleus lies directly in the penetration path, no deviation from the sperms' original course is necessary (B).

Chemistry of Fertilization

The eggs and sperm contain chemicals that are necessary for fertilization. Because they are produced by the gametes and are similar to hormones, the term *gamones* has been applied to them. Those that are found in the sperm are called the *androgamones*, whereas those in the eggs are the *gynogamones*. Two have been found to occur in the sperm, *androgamone I* and *androgamone II*, and two in the egg, *gynogamone I* and *gynogamone II*. Androgamone I is responsible for conserving the sperm's activity. If the sperm were to initially swim as rapidly as possible, all the stored energy would be utilized in too short a period of time. By conserving this activity until the sperm approaches the egg, the probability of fertilization is increased.

Androgamone II, also found in the sperm, dissolves the gelatinous coating around the egg. Androgamone II is indispensable for fertilization, for as long as the gelatinous coating of the egg is intact, the sperm cannot penetrate the egg. Gynogamone I, a hormonelike substance produced by the egg, neutralizes androgamone I. As the sperm approaches the egg, it comes into contact with gynogamone I. Androgamone I now becomes neutralized, and the sperm's activity is consequently increased. Gynogamone II makes the sperm heads sticky. This enables the sperm to stick to the egg surface and allows androgamone II an opportunity to dissolve the gelatinous coating around the egg. A summary of these reactions is given in Figure 3-8.

GAMONE	GAMETE	FUNCTION
Androgamone I	Sperm	Conserves sperm activity
Gynogamone I	Egg	Neutralizes androgamone I, thereby increasing sperm activity
Gynogamone II	Egg	Makes sperm head sticky for easier attachment to egg surface
Androgamone II	Sperm	Dissolves vitelline membrane, allowing sperm entrance

Figure 3-8. Chemistry of fertilization.

Development Subsequent to Fertilization

In all vertebrate eggs, a sequence of developmental stages occurs which is basically similar in all the forms to be studied (Fig. 3-9). Each of these stages will be covered more thoroughly in subsequent chapters. Once the egg is fertilized, it begins to cleave or divide into many smaller cells. This process is called *cleavage.* The end result of the cleavage process is a hollow ball of cells known as the *blastula.*

Following this, a series of cellular movements occur that establish the three germ layers of the embryo, i.e., the *ectoderm* or outside layer, *mesoderm* or middle layer, and *endoderm* or inner layer of cells. This process is *gastrulation*, and when the embryo is in this stage, it is referred to as a *gastrula*. The last of the early embryonic stages is the *neurula*. When the embryo is in this stage, many internal processes are occurring which result in the formation of the *coelomic cavity*, the *notochord*, and the *neural tube*, as well as other primitive organ structures.

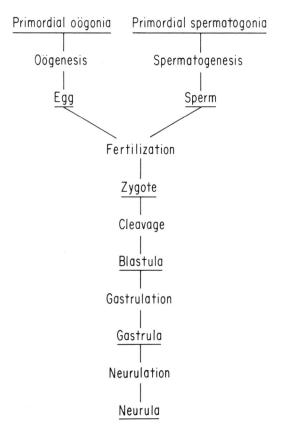

Figure 3-9. Early developmental sequence.

"As thou knowest not what is the way of the spirit, nor how the bones do grow in the womb of her that is with child: even so thou knowest not the works of God who maketh all." Ecclesiastes 11:5.

Chapter 4

Cleavage

Vertebrates are all multicellular animals. When one considers that the fertilized egg is but a single cell, it becomes immediately apparent that some type of cell division must occur before the time of fertilization and the completion of development. The initial phase of this cellular division is called *cleavage*, and the cells resulting from this division are known as embryonic cells or *blastomeres*. In some respects, cleavage might be considered to be a form of mitosis. However, a closer examination reveals that there is a basic difference between the two processes. During normal mitosis, i.e., as it occurs in the adult body, the individual cells undergo a period of growth between successive divisions (Fig. 4-1A). No such interval of blastomere growth occurs in cleavage (Fig. 4-1B). As a result, each succeeding division in the cleavage process produces smaller cells.

Every cell of an adult vertebrate has a *nucleoplasmic ratio*, i.e., a ratio between nuclear and cytoplasmic material, which is specific for the species to which it belongs. Before cleavage begins, the ratio of nuclear to cytoplasmic material is very small. The progressive reduction in size of the individual blastomeres during the cleavage process continues until the specific nucleoplasmic ratio of its species is reached. At this point, the cleavage process ends. As an example of this stabilization of the nucleoplasmic ratio, let us consider the sea urchin egg. Before cleavage, the volumetric ratio of nucleus to cytoplasm is 1:550. At the close of cleavage, it is 1:6, the ratio specific for the mature sea urchin. Once cleavage is completed, normal mitosis, with interdivisional cellular growth, begins. During the entire cleavage process, the embryo neither grows appreciably, nor is its general shape changed.

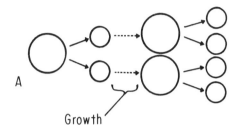

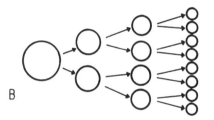

Figure 4-1. Comparison of cleavage and somatic mitosis. During normal mitosis (A) a period of growth occurs after division. During cleavage (B) no such interdivisional growth occurs.

The type of cleavage exhibited by vertebrates depends primarily on the quantity and distribution of yolk in the fertilized egg. This preliminary discussion of cleavage, therefore, must be generalized, and the unique modifications of the cleavage pattern exhibited by specific vertebrates left for the chapters dealing with their individual development. Numerically, early cleavage produces cells at a rate approximating a geometric progression. The first cleavage produces two cells which in turn divide to form four blastomeres. This continues, resulting progressively in 8, 16, and 32 blastomeres. After this, in most vertebrate eggs, the cleavage pattern becomes so irregular that it is impossible to follow it accurately.

Cleavage is actually an external manifestation of an internal process. The single nucleus of the fertilized egg undergoes a mitotic division, and that portion of the process which is commonly called the cytoplasmic division is seen by the observer as a cleavage plane. The classical pattern of these cleavage planes is illustrated in Figure 4-2. The first plane begins at the animal and progresses toward the vegetal pole, separating the egg into two blastomeres. The second

Figure 4-2. Diagrammatic representation of the first five cleavage planes. (A) The first plane begins at the animal and progresses towards the vegetal pole, separating the egg into two blastomeres. (B) The second cleavage plane is also oriented from animal to vegetal pole but at right angles to the first plane. This results in four blastomeres. (C) The third cleavage plane is horizontal, midway between the animal and vegetal poles, at right angles to both planes 1 and 2. This plane separates the four blastomeres into four animal and four vegetal blastomeres. (D) The fourth cleavage plane is, in reality, two separate planes which run from the animal to the vegetal pole. As this double plane transects each of the cells, each hemisphere now contains eight blastomeres. (E) The fifth cleavage plane is again a double plane, each one parallel to plane 3 and at right angles to planes 1, 2, and 4. One occurs in the animal, the other in the vegetal hemisphere. This results in 16 animal and 16 vegetal blastomeres.

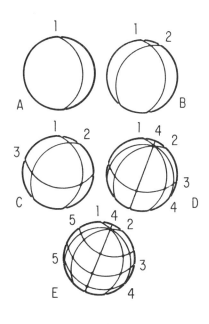

cleavage plane is also oriented from animal to vegetal pole but at right angles to the first plane. This results in four blastomeres. The third cleavage plane is horizontal, midway between the animal and vegetal poles. It is, thus, at right angles to both cleavage planes one and two. This plane separates the four blastomeres into four animal and four vegetal blastomeres. The fourth cleavage plane is, in reality, two separate planes, both of which run from the animal to the vegetal pole. As this double plane transects each of the cells, each hemisphere now contains eight blastomeres. The fifth cleavage plane is again a double plane, each one parallel to plane three and at right angles to planes one, two, and four. One occurs in the animal, the other in the vegetal hemisphere. This results in 16 animal and 16 vegetal blastomeres. This classical pattern occurs only in alecithal eggs, i.e., those that are completely devoid of yolk. Since there are no vertebrate eggs of this type, this basic pattern will be found only in a modified form. The extent of the modification will depend, primarily, on the amount and distribution of the inert yolk material in the egg. The yolk, in vertebrate eggs, is also called the *deutoplasm*. It retards cleavage by presenting a physical barrier to the orderly rearrangement of nuclear material and the formation of the spindle, both of which are so necessary during mitosis.

Three generalized rules of cleavage have been postulated on the basis of yolk content and distribution. These are *Balfour's law*, *Sach's rules*, and *Hertwig's rules*. Balfour's law states that *the rate of cleavage is inversely proportional to the amount of yolk in the egg*. Sach's rules postulate that (1) *cells typically tend to divide into equal parts*, and that (2) *each new plane of division tends to intersect the preceding one at right angles*. Hertwig's rules state that (1) *the typical position of the nucleus is in the center of the protoplasmic mass in which it lies*, and that (2) *the axis of the spindle lies typically in the longest axis of the protoplasmic mass, and division, therefore, cuts this axis transversely.*

The amount of yolk in an egg is the primary factor responsible for the alteration of the cleavage pattern. As was seen in Chapter 2, eggs are classified on the basis of the amount and distribution of yolk present. The yolk in all vertebrate eggs is distributed along a vegetal-animal axis, with the greatest concentration occurring at the vegetal pole. The retardation of development imposed on the egg by this yolk allows us to distinguish two general types of cleavage patterns.

The first is *holoblastic*, indicating that the entire egg cleaves. This type is further subdivided into *holoblastic equal cleavage*, in which the resulting blastomeres are approximately equal in size, and *holoblastic unequal cleavage*, in which the blastomeres of the vegetal hemisphere are larger than those in the animal hemisphere. The equality of blastomere size in holoblastic equal cleavage indicates that either no yolk or only a very slight amount of yolk is present in the egg. Eggs possessing a moderate amount of yolk, concentrated in the vegetal hemisphere, will exhibit holoblastic unequal cleavage. Since yolk inhibits the formation of the mitotic cleavage spindle, the cells in the animal hemisphere will divide faster, and the resulting cells consequently will be smaller than those in the vegetal hemisphere, where yolk has been impeding the progress of cleavage.

A second general type of cleavage is *meroblastic* or *discoidal*, exhibited by those eggs with an extremely large amount of yolk concentrated at the vegetal pole. In this type, the entire egg does not cleave. The cleavage pattern is limited to a small disk of relatively yolk-free cytoplasm at the animal pole. Each of these cleavage types is important enough, and representative of enough vertebrate species, to warrant separate discussion.

Holoblastic Equal Cleavage

This type of cleavage is characteristic of those eggs that have either no yolk (alecithal) or an extremely small amount (meiolecithal). Because there is no impediment to the formation of the spindle or the subsequent cytoplasmic division, the pattern of cleavage is extremely regular. Strictly speaking, holoblastic equal cleavage does not exist in chordate development. However, some eggs, e.g., Amphioxus, possess so little yolk that most embryologists consider them to be alecithal and to exhibit holoblastic equal cleavage. The first cleavage begins at the animal pole and continues around the egg to the vegetal pole, dividing it into two equal blastomeres. The second cleavage also begins at the animal pole but is oriented at right angles to the first cleavage plane. The result of this cleavage is four equal blastomeres. The third cleavage is meridional, that is at right angles to cleavage planes one and two. It is located approximately halfway between the animal and vegetal poles in those eggs that have no yolk, or slightly displaced toward the animal pole in those eggs that have a slight amount of yolk concentrated at the vegetal pole. The result, in the former case, is eight equal blastomeres and, in the latter case, four slightly smaller animal blastomeres and four larger vegetal blastomeres. For our purposes, all these cells will be considered to be equal size. The fourth cleavage is really a double one, each beginning at the animal pole and continuing around the egg to the vegetal pole, at right angles to each other and to cleavage plane three. This results in eight animal and eight vegetal blastomeres. The fifth cleavage plane, the last one that can be followed exactly, is also a double plane. One is located midway between the animal pole and the equator, parallel to the third and perpendicular to cleavage planes one, two, and four. The other is in a similar position in the vegetal hemisphere. The result of this fifth double cleavage plane is 16 animal blastomeres and 16 vegetal blastomeres. Figure 4-3 illustrates the first five cleavage planes in an egg exhibiting holoblastic equal cleavage.

The preceding paragraph has described the first five embryonic stages of a developing egg. Embryologists refer to each of the early stages according to the number of blastomeres they contain. Hence, these eggs have progressed through the 2-celled, 4-celled, 16-celled, and 32-celled stages. Cleavage then continues until the entire surface of the egg is covered with small cells. The egg itself looks like a small raspberry, and the term *morula* is applied by some embryologists to describe this stage.

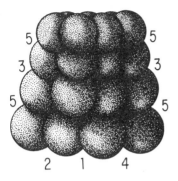

Figure 4-3. Isolecithal egg in a 16-cell stage. Note that although this egg is called "isolecithal," a slight disparity exists between the smaller animal and larger vegetal blastomeres. This is due to the presence of a small amount of yolk in the vegetal hemisphere.

Holoblastic Unequal Cleavage

In eggs with this type of cleavage, the yolk is distributed along a vegetal-animal axis. The yolk is very sparse in the region surrounding the animal pole and increases in concentration as the vegetal pole is approached. Initially, this heavy concentration of yolk in the vegetal hemisphere will, according to Hertwig's rule, cause the fusion nucleus of the uncleaved egg to be dorsally displaced. This dorsal displacement of the nucleus results in a consequent dorsal displacement of the first cleavage spindle. It does not, however, affect the position of the first cleavage plane, for this cuts the axis of the spindle transversely. This plane begins at the animal pole and proceeds around the egg to the vegetal pole. As the cleavage line approaches the vegetal pole, its rate is slowed down. This is an example of Balfour's law. The result of the first cleavage is two equal blastomeres, each containing a heavy concentration of yolk in its vegetal end and very little in the animal hemisphere. The second cleavage likewise is oriented from the animal to the vegetal pole, but it is at right angles to the first cleavage. The result is four equal blastomeres, each of which still has an unequal distribution of yolk within it. The third cleavage is significant, for, because of the heavy concentration of yolk in the vegetal hemisphere, it is dorsally displaced. This plane, being dorsally displaced, creates four smaller animal and four larger vegetal blastomeres. To denote the basic difference in size between the blastomeres of each hemisphere, the term *micromeres* is reserved for the smaller animal blastomeres, and *macromeres* for the larger vegetal blastomeres. The result of the third cleavage plane is, therefore, four animal micromeres and four vegetal macromeres. The fourth cleavage plane is actually a double one, each oriented from animal to vegetal pole. The result of this double cleavage is eight animal micromeres and eight vegetal macromeres. The fifth cleavage plane, again a double one, divides each blastomere transversely, resulting in 16 animal micromeres and 16 vegetal macromeres. The cleavage planes of a holoblastic unequal egg are depicted in Figure 4-4. As development proceeds, the difference in size between the animal and vegetal blastomeres becomes even more noticeable. Numerous internal changes are brought about as a result of this unequal distribution of cell size. These internal changes will be discussed in the next chapter.

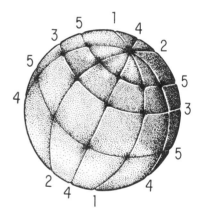

Figure 4-4. Cleavage in a moderately telolecithal egg. As a result of a yolk-inhibited cleavage rate, the blastomeres of the vegetal hemisphere are much larger than those towards the animal pole.

Meroblastic or Discoidal Cleavage

Some eggs, those classified as heavily telolecithal, have such an enormous amount of yolk in the vegetal hemisphere that the active portion of the egg is confined to a small cytoplasmic region at the animal pole. This active area of cytoplasm is the *germinal disk*. Cleavage, instead of involving the entire egg, occurs only in this germinal disk. The first cleavage plane starts in the animal hemisphere and continues toward the lateral portion of the germinal disk. The second cleavage plane occurs at right angles to the first and cuts the germinal disk into four relatively equal, but incomplete, blastomeres. After this, cleavage becomes relatively irregular. Figure 4-5 illustrates the early cleavage pattern in an egg of this type. These early blastomeres are continuous with the yolk on both their lateral and ventral surfaces. This type of cleavage, although differing significantly from the holoblastic equal and holoblastic unequal cleavage, is actually the same process. The basic difference is that it is confined to the germinal disk as a result of the inhibition imposed upon it by the large quantity of yolk.

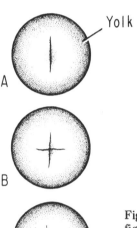

Figure 4-5. Cleavage in a heavily telolecithal egg. The figure above represents only the germinal disk which has been removed from the large mass of underlying yolk. Early cleavage in this type of egg is not complete, since the borders of the early blastomeres are continuous peripherally and ventrally with the underlying yolk.

Significance of Cleavage

The fertilized egg must contain all the factors, or at least the precursors of all the factors, that are responsible for the direction of the complete development of the embryo. These factors must be localized in either the nucleus or the cytoplasm or both. Does cleavage, by dividing both the nucleus and the cytoplasm, also divide the developmental potentials? If it does, each blastomere of the 2-cell stage would have a different set of developmental factors within it. If it does not, each blastomere would be equipotential in terms of its developmental potential. The answer to this question was obtained by removing one of the blastomeres from the 2-cell stage of the frog embryo and watching its development (Fig. 4-6). If the developmental potentials were segregated by the cleavage process, this blas-

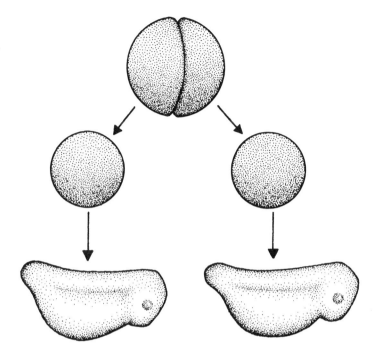

Figure 4-6. Experiment illustrating the totipotency of an isolated blastomere of the 2-cell stage. When an individual blastomere is removed from the 2-cell stage, it develops into a normal, although smaller, larva. This indicates that cleavage has not segregated developmental potential, for if it had, only a partial larva would develop.

tomere, which represents only one-half of the entire egg, would develop at best into a partial larva. Actually, these isolated halves develop into normal, though slightly smaller, larva, indicating that early cleavage in the vertebrate egg does not segregate developmental potentials and that the blastomeres are at this stage equipotential.

Cleavage results in the division of nuclei as well as cytoplasm. Perhaps cleavage acts as an instrument for segregating the developmental potential of the nuclei. To test this hypothesis two workers, Robert Briggs and Thomas King* developed a technique by which they removed the nucleus from a recently fertilized amphibian egg

---

*R. Briggs and T.J. King, 1952, Transportation of living nuclei from blastula cells into enucleated frogs' eggs, Proc. Nat. Acad. Sci. Wash., 38:455-463.

and replaced it with a nucleus taken from later stages of cleavage. Each nucleus which was implanted from late cleavage stages was still capable of directing the egg through complete normal development. This indicated that these nuclei still retained complete capacities for development and that cleavage was not, therefore, an instrument for segregating the developmental potential of cleavage nuclei.

"And ye shall know the truth, and the truth shall make you free." John 8:32.

Chapter **5**

Blastula and Gastrula Formation

Blastulation

Blastulation in an Isolecithal Egg

During the process of cleavage in an isolecithal egg, the individual blastomeres have a tendency to become arranged in a single layer toward the periphery of the developing egg. This creates a centrally located cavity, the *blastocoele.* The embryo, when it is in this stage, is known as a *blastula.* The peripherally located blastomeres become intimately associated with one another, forming a true epithelial layer, the *blastoderm*, which means embryonic skin. Since there is no example of a true isolecithal egg among the chordates, a slight discrepancy in cell size exists, due to yolk inhibition of the vegetal cells. The blastomeres located in the vegetal portion of the blastoderm are slightly larger than those found in the animal region. Because of this slight size discrepancy, an animal-vegetal polarity can be discerned in the blastula stage of an isolecithal egg. An example of this type of blastula is depicted in Figure 5-1.

Blastulation in a Moderately Telolecithal Egg

Throughout segmentation, the moderate amount of yolk in the vegetal hemisphere has, as a result of its inhibition of development,

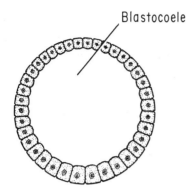
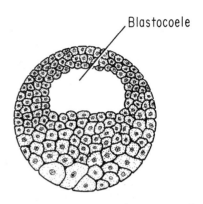

Figure 5-1. Blastula of an isolecithal egg. Notice the slight size discrepancy between the vegetal and animal blastomeres. This is due to a slight amount of yolk in the vegetal hemisphere.

Figure 5-2. Blastula of a moderately telolecithal egg. The dorsal displacement of the blastocoele is due, primarily, to the yolk-inhibited mitotic divisions of the vegetal blastomeres.

produced two modifications of the basic pattern exhibited by the isolecithal egg. First, there is a displacement of all the horizontal cleavages toward the animal pole. Second, the yolk has retarded the rate of cleavage in the vegetal hemisphere. The combined result of both these developmental modifications is an eccentric blastula. The blastocoele is dorsally displaced, and the vegetal blastomeres, which are much larger than those of the animal hemisphere, contain within them a considerable amount of nutritive yolk. The blastula of a moderately telolecithal egg can be seen in Figure 5-2.

Blastulation in a Heavily Telolecithal Egg

In this type of egg, both cleavage and blastulation are restricted to the germinal disk. As cleavage progresses, the cells in the central portion of the disk become separated from the surface of the yolk. The metabolism of the underlying yolk, by the centrally located blastomeres, is responsible for this apparent "lifting" of the cells. Viewed from the surface, two relatively distinct areas can be distinguished. The *area pellucida* is the clearer central area. This is surrounded by the more dense *area opaca*. In this latter area, the ventral and lateral surfaces of the blastomeres are in intimate contact with the yolk. These two areas appear as seen in Figure 5-3. A new layer of cells now appears on the surface of the yolk, below the original layer. After this layer forms, the top layer is referred to as the *epiblast*, whereas this new layer on the surface of the yolk is called the *hypoblast* (Fig. 5-4).

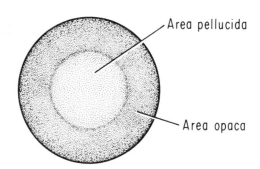

Figure 5-3. Blastodisk of a heavily telolecithal egg showing the two distinct areas. The area pellucida appears lighter because the cells of this area have "lifted" from the yolk surface. The blastomeres in the area opaca still retain an intimate contact with the yolk.

Figure 5-4. Sagittal view of the blastodisk of a heavily telolecithal egg. Utilization of the yolk by the blastomeres of the area pellucida produces a subgerminal cavity.

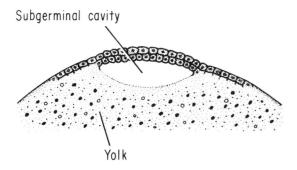

The origin of this hypoblast is not yet completely understood. The real origin is obscured partly because of the impossibility of observing the action of movement of cells in stained preparations. Three general theories have been proposed regarding the hypoblast origin. The first suggests that the thickened blastodisk delaminates to form the hypoblast. The process of *delamination* is one in which a cellular complex splits, forming two separate layers. This process is graphically depicted in Figure 5-5. A second general hypothesis suggests that the posterior portion of the blastodisk undergoes a type of

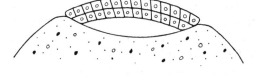

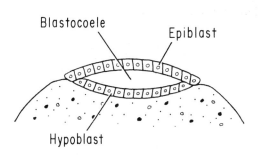

Figure 5-5. Delamination. The process of delamination is essentially a splitting phenomenon resulting in an upper epiblast and a lower hypoblast. The cavity between these two layers is the blastocoele.

involution to form the hypoblast. The term *involution* is reserved for that process by which a layer of cells folds under itself while continuing to grow, thereby forming a new layer beneath the original. The involution process is depicted in Figure 5-6. A third hypothesis suggests that the individual cells of the blastodisk leave the original blastoderm and infiltrate or migrate down to the surface of the yolk. When a number of these cells have reached the yolk surface, they

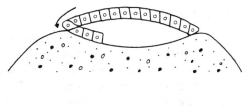

Figure 5-6. The process of involution. The blastomeres of the original blastoderm turn under at the posterior edge of the blastodisk and grow inward to form a new layer. When this process is completed, the involuted layer is referred to as the hypoblast, whereas the original blastoderm is now the epiblast. The cavity between these two new layers is the blastocoele.

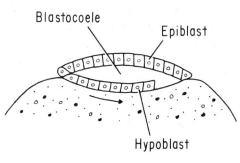

become intimately associated with one another to form the hypoblast. The process of *infiltration* is represented in Figure 5-7. Regardless of which of these mechanisms is involved in the formation of both the epiblast and hypoblast, the end result is the same. The cavity between the epiblast and hypoblast is now referred to as the *blastocoele*. This is, in all respects, homologous to the blastocoeles of both the isolecithal and moderately telolecithal eggs.

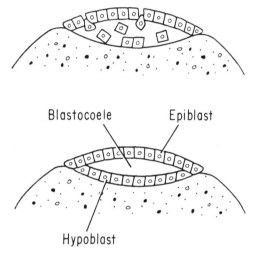

Figure 5-7. Infiltration. Individual cells of the blastodisk leave the original blastoderm and infiltrate or migrate down to the surface of the yolk. When these cells reach the yolk surface, they become intimately associated with one another to form the hypoblast.

Gastrulation

Gastrulation is that process which follows immediately after the embryo reaches the blastula stage. During gastrulation, the cells undergo a series of movements or shifts to form the three definitive germ layers of the adult organism. These three germ layers, the *ectoderm*, *endoderm*, and *mesoderm*, are already localized in their presumptive form in the original blastoderm. It is possible to stain portions of the blastula with nontoxic dyes. These stains remain localized and allow an investigator to follow cellular movement. This technique, *vital staining*, is used extensively by embryologists. By utilizing this staining technique, these presumptive areas can be localized, and their movements actually can be followed during the gastrulation process. Although the three germ layers are established during gastrulation in all vertebrates, the mechanism differs. This is due, primarily, to the morphological uniqueness of the three blastula types just discussed. The description that follows, however, must at this time be generalized, for developmental differences exist even among vertebrates with similar blastulae. A more complete discussion of gastrulation will be undertaken as each vertebrate type is studied.

Gastrulation
in an Isolecithal Egg

In an isolecithal egg, the blastocoele is, for all practical purposes, centrally located, although a slight size polarity of the blastomeres is exhibited along the animal-vegetal axis. The beginning of gastrulation is indicated by a flattening of the vegetal hemisphere. This area of the blastula then begins to push inward, a process known as *invagination* (Fig. 5-8). The morphological result of this invagination is a cuplike structure, composed of two layers of cells, an outer ectoderm and an inner *mesendoderm*, so called because it contains the presumptive material for both the mesoderm and endoderm. The new cavity that is formed is the *archenteron*. The opening into this cavity is the *blastopore*, which can be considered analogous to a

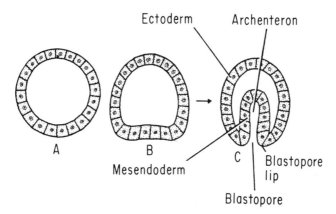

Figure 5-8. Gastrulation in an isolecithal egg. (A) In an isolecithal egg, the blastocoele is, for all practical purposes, centrally located, although a slight size gradient of the blastomeres is exhibited along the animal-vegetal axis. (B) The vegetal hemisphere flattens and begins to push inward, a process known as invagination. (C) Continued invagination produces a cuplike structure composed of two layers of cells, an outer ectoderm and an inner mesendoderm. The new cavity is the archenteron. The opening into this cavity is the blastopore. The edges of the blastopore are the lips. (See also Fig. 5-9).

mouth. Developing this analogy further, the edges of the blastopore are its *lips*. We can speak, therefore, of the *dorsal, ventral,* and *lateral lips of the blastopore* (Fig. 5-9). As invagination proceeds, the original blastocoele is nearly obliterated by the inpushing layer of cells. The lips of the blastopore, at first forming a circular opening, gradually approach one another until the original blastopore is reduced to a mere slit in the outer surface. When the embryo is in this condition, it is referred to as a *gastrula*.

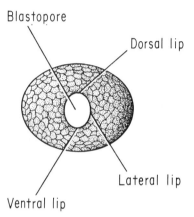

Figure 5-9. External view of the gastrula showing the opening into the archenteron (blastopore) surrounded by the dorsal, lateral, and ventral lips of the blastopore.

Gastrulation in a Moderately Telolecithal Egg

In a moderately telolecithal egg, the blastocoele is displaced toward the animal pole. The large blastomeres of the vegetal hemisphere preclude any type of invagination by the vegetal half of the egg. Consequently, a different type of cellular movement occurs during the gastrulation process of a moderately telolecithal egg. This cellular movement is called *involution* and can best be described by utilizing a series of figures. Figure 5-10 illustrates this process. Figure 5-10A is a midsagittal section through the blastula of a moderately telolecithal egg. Numbers 1 through 9 represent specific regions of the blastoderm of the animal hemisphere, with 5 occurring at the animal pole. From our discussion of cleavage, it will be remembered that the blastomeres of the animal hemisphere are dividing at a more rapid rate than those near the vegetal pole, since the cells in the vegetal hemisphere still contain large quantities of mitosis-inhibiting yolk. Now if we analogize this blastula with an orange, over which

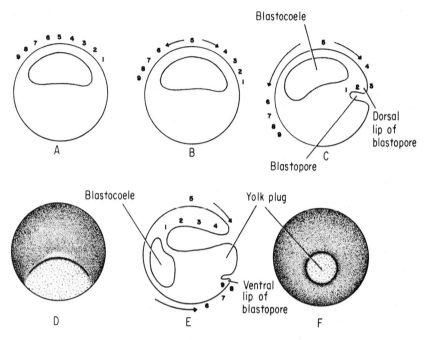

Figure 5-10. Gastrulation in a moderately telolecithal egg. (A) Midsagittal section through the blastula of a moderately telolecithal egg. Numbers 1 through 9 mark specific regions of the blastoderm of the animal hemisphere, with 5 occurring at the animal pole. (B) As a result of intensive mitotic activity, the animal blastomeres begin to migrate toward the vegetal pole, a process known as epiboly (See Fig. 5-11). (C) Formation of the dorsal lip of the blastopore. The region illustrated by 1 and 2 has already started to "round the corner" and grow inward. This process is involution (See also Fig. 5-6). (D) External view of the early gastrula shown in section in C. The darker line separating the animal blastomeres from those of the vegetal hemisphere represents the dorsal lip of the blastopore, the site of involution. (E) As development of the gastrula proceeds, more and more of the cells (1 through 4) will involute, until eventually the configuration shown is reached. The blastocoele has been nearly obliterated by the evergrowing archenteron. The cells represented by 6 through 9 are not stopped in their progress toward the vegetal pole. Instead, they continue around the egg until they approach the blastopore from the ventral side. Here they also involute, forming the ventral lip of the blastopore. (F) External view of the late gastrula, commonly called the yolk-plug stage. The large yolk laden vegetal cells can be seen through the newly formed blastopore.

syrup is being poured, we will have a relationship which is fairly accurate. If the syrup is poured over the orange, it will flow down and around the sphere in all directions (Fig. 5-11). This is practically what happens in the case of the *blastula*, for the cells in the region of 5 are reproducing at such a fast rate, relative to the vegetal cells, that they must either accumulate or "flow down" over the ventral surface of the egg. This flowing of cells is referred to as *epiboly* and is very instrumental in the formation of the gastrula. Figure 5-10B catches these cells as they are beginning to flow around the surface of the egg. At one part of the egg's surface the cells cease their downward movement and begin to flow into the blastula itself. In Figure 5-10C the region illustrated by 1 and 2 has already started to "round the corner" and grow inward. When a layer of cells grows into a cavity rather than simply pushing into it, as is the case in invagination, the process is called *involution*. This combination of epiboly and subsequent involution creates a number of new morphological structures. The new cavity which is beginning to form is the *gastrocoele* or archenteron. The opening into this archenteron is the blastopore. Figure 5-10D shows this early gastrula at approximately the same stage as is seen in Figure 5-10C, but views it as it would be seen by looking into the blastopore. As development of the gastrula proceeds, more and more of the cells indicated by 1 through 4 will involute, until eventually the configuration shown in Figure 5-10E is reached. The blastocoele has been nearly obliterated by the ever-

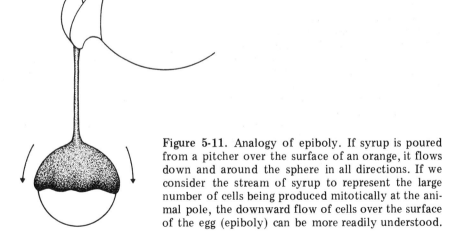

Figure 5-11. Analogy of epiboly. If syrup is poured from a pitcher over the surface of an orange, it flows down and around the sphere in all directions. If we consider the stream of syrup to represent the large number of cells being produced mitotically at the animal pole, the downward flow of cells over the surface of the egg (epiboly) can be more readily understood.

growing archenteron. The cells represented by 6 to 9, however, are not stopped in their progression toward the vegetal pole. Instead, they continue around the egg until they also approach the blastopore from the ventral side (Figs. 5-10C and E). Here they also involute, forming the ventral lip of the blastopore. As a result of these cellular movements, the entire portion is occupied by the blastopore. Through this can be seen the large yolk-laden blastomeres of the original ventral surface of the blastoderm. It is for this reason that this stage is sometimes referred to as the yolk-plug stage (Fig. 5-10F). A cross section through the embryo at this stage is depicted in Figure 5-12. The floor of the archenteron is endoderm, whereas the roof is mesendoderm, that is, presumptive mesoderm and endoderm. The outside of the gastrula is, of course, ectoderm. The further development of this gastrula will be discussed in Chapter 6.

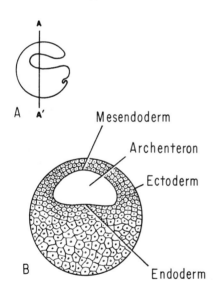

Figure 5-12. Cross section of the yolk-plug stage. If the yolk-plug stage (A) is sectioned along A-A', the configuration shown in B will be seen. The roof of the archenteron is mesendoderm, and the cells on the surface of the embryo are ectoderm.

Gastrulation in a Heavily Telolecithal Egg

Gastrulation in a heavily telolecithal egg can be divided into two major phases: early gastrulation, which includes all the events leading up to the major cellular movements, and gastrulation proper, which establishes the three definitive germ layers. The active part of the blastoderm in early gastrulation is the area pellucida, specifically, the posterior two-thirds of that area. Both medially and posteriorly directed movements of the cells lying in the lateral and anterior regions of the area pellucida result in a thickened accumulation of cells along the midline of this region. This thickening is called the *primitive streak* (Figs. 5-13 and 5-14). The center of the primitive

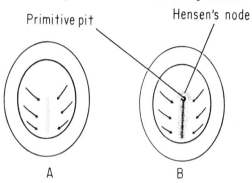

Figure 5-13. Formation of the primitive streak. (A) Cells lying in the anterior and lateral region of the area pellucida converge toward the midline. This thickened area of cells, seen also in Figure 5-14, is called the primitive streak. (B) Anteriorly a thickened area, Hensen's node, and a depression, the primitive pit, can be seen. This area, is homologous to the dorsal lip of the blastopore in the early gastrula of a moderately telolecithal egg.

Figure 5-14. Cross section through the primitive streak. The involuting cells can be seen beneath the primitive streak in intimate contact with the hypoblast. They are also growing out laterally to form the mesoderm.

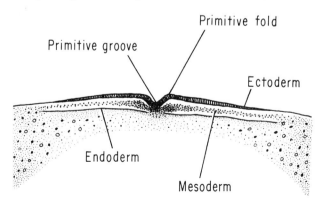

streak is a depression, the *primitive groove*, whereas the edges of the primitive streak are raised and are called the *primitive folds*. Anteriorly, the primitive groove is even more depressed. This depression is the *primitive pit*. The raised area around the primitive pit is *Hensen's node*. This movement of cells from the anterior and lateral regions toward the primitive streak is a form of epiboly, but as the cells converge toward the central area, the process is referred to specifically as *convergence*. While this developmental process is occurring in the epiblast, the hypoblast continues to fill in across the surface of the yolk, establishing a definitive endoderm. When the embryo is in this condition, the major gastrulation movements begin.

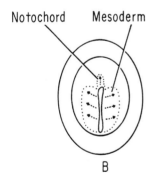

Figure 5-15. Gastrulation in a heavily telolecithal egg. (A) The converging cells move toward the primitive streak. Here they move from the surface of the epiblast to migrate downward until they come into intimate contact with the hypoblast. (B) From these involuting cells a new layer, the mesoderm, is formed. This grows out laterally and anteriorly between the epiblast and hypoblast. Laterally, this layer of cells forms the mesoderm, whereas the anteriorly directed growth is the head process or notochord.

The primitive streak is homologous to the blastopore of the moderately telolecithal egg. If this homology can be visualized, the movements of the cells throughout the gastrulation process in this type of egg will be more readily understood. Figure 5-15 illustrates the process of gastrulation in this egg type. The converging cells move toward the primitive streak. Here they move from the surface of the epiblast and migrate downward until they come into intimate contact with the hypoblast. From these cells a new layer, the mesoderm, is formed. This grows out laterally and anteriorly between the epiblast and hypoblast. Laterally, this layer of cells forms the mesoderm, whereas the anteriorly directed growth is the *head process* or *notochord.* The development of these structures will be followed more thoroughly in Chapter 6.

"For God giveth to a man that is good in his sight wisdom, and knowledge, and joy." Ecclesiastes 2:26.

Chapter 6

Neurulation

Embryological development is a continuous process. Although it is frequently studied in stages, as has already been done in this book, it must be understood that these stages are man-made. This artificial staging is similar to stopping a motion picture, in order to study a particular frame. This breaking down of a phenomenon as complex as embryological development into definitive periods has its shortcomings, for many morphological processes overlap one or more stages. For this reason it is difficult to isolate clearly one stage from the next. The current chapter is concerned primarily with the formation of the neural tube, the primordium of the central nervous system. During this same period, however, other developmental processes are occurring, processes that started during gastrulation. These include, among others, the development of the mesoderm, coelomic cavity, and primitive digestive tract (gut). All of these will, therefore, be studied in this chapter, although some would argue, legitimately, that the development of these structures does not properly belong in a chapter on neurulation.

Mesoderm and Notochord Formation in an Isolecithal Egg

After gastrulation, the embryo is in a diploblastic form, with an outer ectoderm and an inner cavity lined with mesendoderm. Shortly after this, the embryo begins to rotate in such a manner that the original animal pole becomes the anterior end of the animal, and the vegetal pole, the site of the blastopore, becomes the posterior end (Fig. 6-1). A cross section through an embryo at this stage would reveal the structures seen in Figure 6-2. The roof of the archenteron

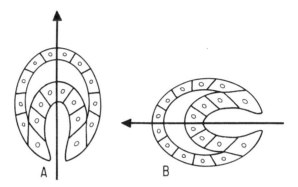

Figure 6-1. Rotation of an isolecithal egg. (A) The animal pole (head of the arrow) is uppermost. (B) As a result of egg rotation, the animal pole becomes the future anterior end of the larva. Consequently, the blastopore is located at the future posterior end.

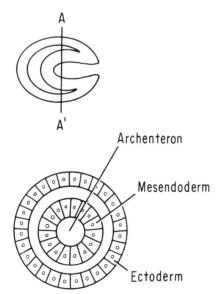

Figure 6-2. Cross section through a gastrula of an isolecithal egg cut along plane A-A′. The embryo at this stage is diploblastic, with an outer ectoderm and an inner mesendoderm. The central lumen is the archenteron, the primordium of the digestive system.

is the *chordamesoderm*. This name is applied to it because it contains both presumptive notochordal and mesodermal materials. The floor of the archenteron is endoderm. A pair of grooves appear laterally in the roof of the archenteron. These grooves begin to deepen, forming two lateral pouches and pinching off a central portion. This central portion is the future notochord, whereas the two pouches that are forming laterally represent the mesoderm. This process continues (Fig. 6-3) until the three portions separate from each other and from the ventrally located endoderm. The dorsal ends of the endoderm, once freed from the chordamesoderm, fuse to one another, forming for the first time a gut cavity completely lined with endoderm. Each of the mesodermal pouches grows laterally and ventrally until they meet one another beneath the gut. The fused portions then disintegrate, creating a mesodermal complex composed of an inner and outer layer of mesoderm and an enclosed cavity. The inner layer of mesoderm, that next to the endoderm, is called *splanchnic meso-*

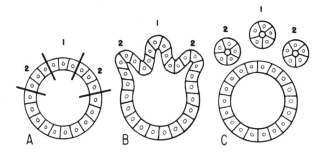

Figure 6-3. Cross section through the archenteron of an isolecithal egg. (A) The dorsal half of the archenteron indicated by 1 and 2 is the chordamesoderm. It contains both presumptive notochordal and mesodermal materials. The ventral half of the archenteron is endoderm. (B) A pair of grooves (2-2) appears laterally in the roof of the archenteron. These grooves begin to deepen, forming two lateral pouches (2-2) and pinching off a central portion (1). (C) This process continues until the three portions separate from each other and from the ventrally located endoderm. The central pouch (1) is the future notochord, whereas the two lateral pouches (2-2) represent the mesoderm. The dorsal ends of the endoderm, once freed from the chordamesoderm, fuse to one another to form a gut cavity completely lined with endoderm.

derm, whereas the outer layer, that next to the ectoderm, is the *somatic mesoderm*. The cavity between the somatic and splanchnic mesoderm is the *coelome*. This proximity between splanchnic mesoderm and endoderm, and somatic mesoderm and ectoderm, is developmentally significant. Many of the embryonic structures which develop result from infoldings or outfoldings of these combined layers. So often is this the case that a single term is used to designate each of these dual layers. The splanchnic mesoderm and endoderm are called collectively the *splanchnopleure*, whereas the term *somatopleure* is reserved for the combination of somatic mesoderm and ectoderm. Due to their method of formation, the notochord, mesoderm, and coelomic cavity extend nearly the entire length of the body.

Neurulation in an Isolecithal Egg

The ectoderm immediately above the notochord thickens and is referred to as the *medullary plate*. It is this layer of tissue that will eventually give rise to the *neural tube*. The ectoderm on each side of the medullary plate begins to migrate medially. This migration begins in the region of the blastopore and proceeds anteriorly, giving the impression of a closing zipper (Figs. 6-4 and 6-5). These medially and anteriorly directed ectodermal growths, the *neural ridges*, eventually meet one another in the midline, forming a completely closed roof over the medullary plate. The resulting cavity remains opened at its

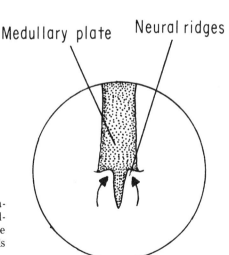

Figure 6-4. Anterior and medial migration of the ectoderm over the medullary plate. This migration begins in the region of the blastopore and proceeds anteriorly as a pair of neural ridges.

anterior end. This opening is the *anterior neuropore* (Fig. 6-6). The medullary plate subsequently folds up to form the *neural tube* (Fig. 6-7). Posteriorly, the overgrowth of the ectoderm has completely obliterated the original blastopore. However, a connection now exists between the neural canal and the primitive gut. This new canal is the *neurenteric canal* (Fig. 6-8).

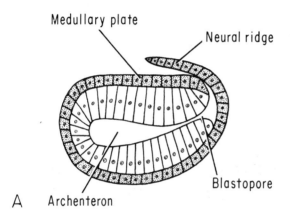

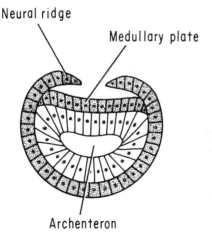

Figure 6-5. Migration of the neural ridges. (A) As the neural ridges migrate anteriorly, they also approach one another in the midline to form a complete roof over the medullary plate. (B) The migrating ectoderm completely closes over the original blastopore.

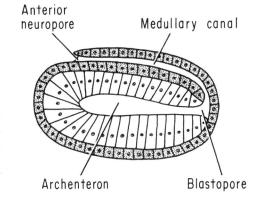

Figure 6-6. Anterior neuropore. The cavity over the medullary plate resulting from the migration of the neural ridges remains open at its anterior end. This opening is the anterior neuropore.

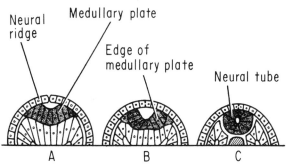

Figure 6-7. Formation of the neural tube. (A) The neural ridges fuse in the midline to form a complete roof over the medullary plate. The edges of the plate begin to elevate. (B) The medullary plate continues to round up, forming the complete neural tube seen in C.

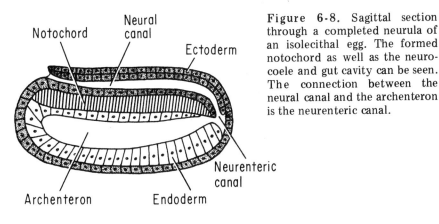

Figure 6-8. Sagittal section through a completed neurula of an isolecithal egg. The formed notochord as well as the neurocoele and gut cavity can be seen. The connection between the neural canal and the archenteron is the neurenteric canal.

Mesoderm and Notochord Formation in a Moderately Telolecithal Egg

It was indicated earlier that the roof of the archenteron is mesendoderm. This thickened archenteric roof undergoes delamination which results in an upper and lower layer of cells. The upper portion of the delaminate is chordamesoderm, for subsequent devel-

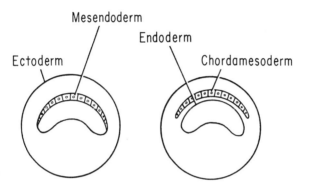

Figure 6-9. Delamination of the archenteric roof in a moderately telolecithal egg. The thickened archenteric roof delaminates to form an upper chordamesoderm and a lower endoderm.

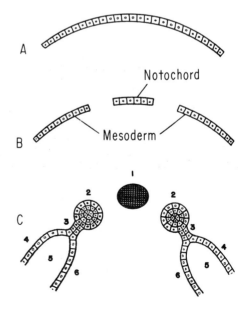

Figure 6-10. Differentiation of the chordamesoderm. (A) The chordamesoderm shortly after delamination. (B) The chordamesoderm divides into a central notochord and two lateral mesodermal sections. (C) The mesoderm itself soon begins to differentiate into three distinct regions. The most medial is the somite (2). This is followed laterally by the intermediate mesoderm (3). A split occurs in the lateral mesoderm dividing it into somatic (4) and splanchnic (6) layers which surround the coelomic cavity (5). The notochord (1) also differentiates.

opment reveals that it gives rise to both the notochord and mesoderm. The lower portion of the delaminate is endoderm and represents the definitive roof of the primitive gut. These early changes are shown in Figure 6-9. The chordamesoderm now divides into three distinct sections, a central notochord, and two lateral mesodermal divisions (Fig. 6-10). The mesoderm itself soon begins to differentiate into three distinct regions. The most medial is the *somite.* This is followed laterally by the *intermediate mesoderm.* Most lateral is the mesoderm proper. The terms *epimere, mesomere,* and *hypomere,* respectively, are often used to designate these regions. A split occurs in the lateral mesoderm (hypomere) as it begins to grow ventrad to meet its mate from the other side. This split is similar to that already seen in the isolecithal egg and represents a division of the lateral mesoderm into the somatic and splanchnic mesoderm. The cavity formed between the two layers is the coelome.

Neurulation in a Moderately Telolecithal Egg

The notochord, because of its method of formation, runs nearly the entire length of the animal in an anterior-posterior direction. It is located immediately beneath the ectoderm which will give rise to the neural tube. This ectoderm is the medullary plate. The medullary plate begins to thicken, and soon the edges start to elevate. These edges are the *neural folds* (Fig. 6-11). The neural folds approach one

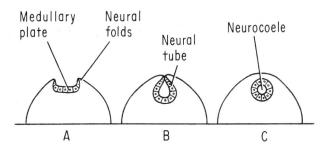

Figure 6-11. Formation of the neural tube in a moderately telolecithal egg. (A) The edges of the medullary plate (the neural folds) begin to elevate. (B) The neural folds approach one another, forming a complete tube. (C) The completed neural tube lies beneath the original ectoderm. The new cavity of the neural tube is the neurocoele.

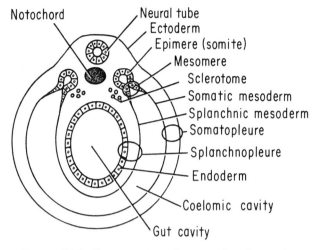

Figure 6-12. Cross section of a neurula of a moderately telolecithal egg.

another in the midline, and as they do, the connection between them dissolves, leaving a neural tube completely enclosed beneath the overlying ectoderm. When the neural tube is completed, the embryo is seen as pictured in Figure 6-12.

Mesoderm and Notochord Formation in a Heavily Telolecithal Egg

The involuting epiblast begins to grow between the epiblast and hypoblast in both anterior and lateral directions. The anteriorly directed portion is the head process or notochord. This lies under what will become the body of the embryo. The lateral expansion of the involuting material forms the mesoderm. Figure 6-13 illustrates the location of this invaginating material after gastrulation is well underway. A cross section through the primitive streak region would reveal that the mesoderm is split and looks very similar to that which has already been studied in both the isolecithal and moderately telolecithal eggs. This mesoderm is divided into a medially located somite (epimere), an intermediate mesoderm (mesomere), and a lateral mesoderm (hypomere). This lateral mesoderm is separated into a somatic and splanchnic layer which encloses the coelomic cavity (Fig. 6-14).

Figure 6-13. Late gastrula of a heavily telolecithal egg. The laterally directed growths are mesoderm, whereas the anterior one is the head process or notochord.

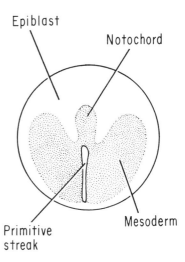

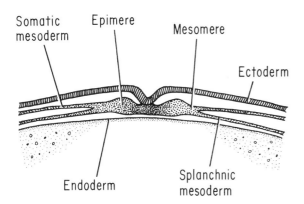

Figure 6-14. Cross section through a late primitive streak stage. The involuting material differentiates into a medial epimere, intermediate mesomere, and lateral hypomere. The latter is further subdivided into somatic and splanchnic layers.

Neurulation in a Heavily Telolecithal Egg

The epiblast, now more correctly called the ectoderm, lying immediately above the notochord soon undergoes a change similar to that already seen in the moderately telolecithal egg. It thickens to form the medullary plate. This in turn folds up to form the neural tube. A cross section through the embryo at the completion of these movements is seen in Figure 6-15.

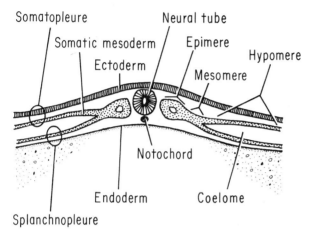

Figure 6-15. Cross section through the neurala of a heavily telolecithal egg.

Student Review

Throughout these first six chapters, an attempt has been made to introduce you to the basic terms, processes, and tenets of embryology. A knowledge of the material contained within these chapters is essential before turning to the far more complex study of the development of specific vertebrate types. For this reason, a list of terms is presented below. You should be able to describe the process, identify the structure, or define the terms. If some still escape you, it is recommended that you return to the earlier pages on which they are defined in order to familiarize yourself with them. The remainder of the book will be more comprehensive if you master these terms before proceeding with Chapter 7.

acrosome
alecithal egg
allantois
amnion
androgamones
animal pole
archenteron
area opaca
area pellucida
artificial parthenogenesis
axial filament
Balfour's law
biogenetic law
blastocoele
blastoderm
blastomeres
blastopore
blastula
chemical embryology
chiasmata
chordamesoderm
chorion
chromatid
chromonemata
chromosomal aberrations
chromosomes
cleavage
cleavage plane
cleidoic egg
coelomic cavity
comparative embryology
convergence
copulation path
cytoplasm
delamination
deletions
descriptive embryology
deutoplasm
developmental physiology
diad
diakinesis
diploblastic

diploid number
diplotene
discoidal cleavage
distal centrosome
dorsal lip of blastopore
ectoderm
egg
embryology
endoderm
epiblast
epiboly
epididymal ducts
epimere
equatorial plate
estrogen
experimental embryology
external fertilization
female pronucleus
fertilization
fertilization cone
fertilization membrane
first polar body
foetal pig
follicular cells
gametes
gametogenesis
gamones
gastrocoele
gastrula
gastrulation
genes
germinal disk
Golgi apparatus
graafian follicle
gut
gynogamones
haploid number
heavily telolecithal egg
Hensen's node
Hertwig's rules
holoblastic equal cleavage
holoblastic unequal cleavage

homologous chromosomes
hypoblast
hypomere
infiltration
intercalary deletion
intercalary translocation
internal fertilization
invagination
inversions
involution
isolecithal egg
lateral lip of blastopore
leptotene
macromeres
male pronucleus
medullary plate
meiolecithal egg
meiosis
meroblastic cleavage
mesectoderm
mesendoderm
mesoderm
mesolecithal egg
mesomere
metamorphosis
metaphase plate
micromeres
mitochondria
mitosis
moderately telolecithal egg
monad
monospermy
mutagenic agent
neural folds
neural ridges
neural tube
neurenteric canal
neuropore
neurula
neurulation
notochord
nucleoplasmic ratio

ontogenetic hypertrophy
ontogeny
oögenesis
oögonia
oötid
ostium
ova
ovaries
oviduct
oviparous
ovoviviparous
ovulation
pachytene
parthenogenesis
penetration path
perivitelline space
phylogeny
placenta
plasma membrane
point mutation
polylecithal egg
polyspermy
position effect
primary oöcytes
primary spermatocytes
primitive folds
primitive groove
primitive pit
primitive streak
primordial germ cells
pronuclei
proximal centrosome
reduction division
rete testes
Sach's rules
scrotal sac
second polar body
secondary oöcytes
secondary spermatocytes
seminiferous tubules
somatic mesoderm
somatopleure

sperm
sperm head
sperm neck
sperm nucleus
sperm tail
spermatids
spermatogenesis
spermatogonia
splanchnic mesoderm
splanchnopleure
synapsis
synaptene
terminal deletion
terminal translocation

testes
tetrad
translocation
vegetal-animal gradient
vegetal pole
ventral lip of blastopore
vital staining
vitelline membrane
viviparous
yolk
yolk plug
yolk sac
zygotene

PART **II**

SPECIFIC EMBRYOLOGY

"And God said, Let the waters bring forth abundantly the moving creature that hath life, and fowl that may fly above the earth in the open firmament of heaven. And God created great whales, and every living creature that moveth, which the waters brought forth abundantly, after their kind, and every winged fowl after his kind: and God saw that it was good. And God blessed them, saying, Be fruitful, and multiply, and fill the waters in the seas, and let fowl multiply in the earth. And the evening and the morning were the fifth day.

And God said, Let the earth bring forth the living creature after his kind, cattle, and creeping thing, and beast of the earth after his kind: and it was so. And God made the beast of the earth after his kind, and cattle after their kind, and every thing that creepeth upon the earth after his kind: and God saw that it was good.

And God said, Let us make man in our image, after our likeness: and let them have dominion over the fish of the sea, and over the fowl of the air, and over the cattle, and over all the earth, and over every creeping thing that creepeth upon the earth. So God created man in his own image, in the image of God created he him; male and female created he them. And God blessed them, and God said unto them, Be fruitful, and multiply, and replenish the earth, and subdue it: and have dominion over the fish of the sea, and over the fowl of the air, and over every living thing that moveth upon the earth." Genesis 1:20-28.

"The earth is the Lord's, and the fulness thereof; the world, and they that dwell therein." Psalms 24:1.

Chapter 7

The Development of Amphioxus

The Amphioxus is a small fishlike animal which lives in the sand of the seashore just below low tide. They are not common animals, and are represented by only a few genera found in certain parts of the world in the tropics and subtropics. It is a member of the Phylum Chordata but it is not a vertebrate. It is included in this textbook because its embryology does exhibit many of the basic features of vertebrate development.

Egg Morphology

The egg of Amphioxus measures approximately 0.10 mm in diameter. Its nucleus is exceptionally large, having a diameter nearly half that of the egg itself. A definite yolk polarity along the animal-vegetal axis is exhibited. The cortex of the egg contains a type of vacuolated protoplasm, the cortical cytoplasm, which is free of yolk. The egg nucleus lies in contact with this layer at a point near the animal pole, and consequently the meiotic divisions occur in this location.

Prior to leaving the body of the female, the developing egg goes through the first phases of meiosis. The first polar body is liberated from the egg surface and is usually lost. The meiotic process ceases temporarily at metaphase II. In this arrested condition, the egg is

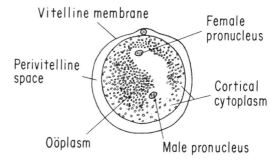

Figure 7-1. The Amphioxus egg. After sperm penetration, but prior to the fusion of the male and female pronuclei, the second polar body can be seen in the perivitelline space. This space is created when the vitelline membrane lifts from the surface of the egg. Since this polar body remains in its position of extrusion, it serves as a marker of the original animal pole.

shed into the water. Contact with the seawater causes the surface membrane (vitelline membrane) to lift from the egg surface as the fertilization membrane. When a sperm enters the egg, it acts as a stimulus, and the meiotic process continues. The second polar body is then extruded from the egg near the animal pole. Here it is trapped in the perivitelline space and serves as a natural marker of the animal pole. This marking device will be used in the description of later stages. The Amphioxus egg, after sperm penetration but prior to the fusion of the male and female pronuclei, can be seen in Figure 7-1.

Fertilization

Fertilization in Amphioxus is external. The sperm are shed directly into the water and swim to the eggs. One or more enter each egg near the vegetal pole. Although more than one sperm frequently enter the egg cytoplasm, all degenerate, except the one that fuses with the female pronucleus. The site of sperm entrance furnishes, together with the animal and vegetal pole, three points of reference which establish the median plane of the future embryo. The area of the egg in which the point of sperm entrance is located becomes the future posterior end of the embryo. These planes are illustrated in Figure 7-2. As a result of this established symmetry, the egg, although still spherical, can be discussed in the anatomical terms of the adult. Hence there is an anterior, posterior, dorsal, ventral, left, and right portion of the egg.

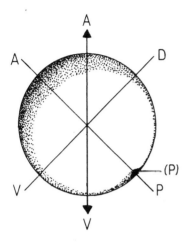

Figure 7-2. Planes of symmetry established by fertilization. The midsagittal plane passes through the animal pole and the point of sperm entrance (P). The animal-vegetal axis is indicated by the line A-V, the anterior posterior axis by A-P, and the dorsal ventral axis by D-V.

Once the meiotic process is completed, the female pronucleus (egg nucleus) and male pronucleus (sperm nucleus) migrate toward one another and fuse in the posterior region of the egg, slightly above the equator. This fusion nucleus soon undergoes its first mitotic division. The axis of the first cleavage spindle is oriented perpendicular to the primary animal-vegetal axis of the egg (Fig. 7-3).

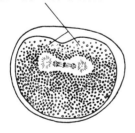

Figure 7-3. Fertilized egg undergoing the first cleavage. Note that the line of demarcation (the first cleavage plane) between the first two blastomeres is oriented ar right angles to the first mitotic spindle.

Precleavage Egg Symmetry

Two distinct areas of the egg have already been seen. The first is the cortical layer of cytoplasm containing the vacuoles. The second is the cytoplasm containing the yolk. As the fusion nucleus begins its first mitotic division, it becomes surrounded by a clear layer of cytoplasm referred to as hyaloplasm. This hyaloplasm spreads from the animal pole toward the vegetal hemisphere. Still another change occurs near the egg surface. As the sperm moves toward the female pronucleus, it drags along with it some of the cortical cytoplasm. The corticoplasm remaining on the egg surface moves around the egg until it forms a crescent of material near the sperm entrance point (Fig. 7-4). This is called the mesodermal crescent because it will give rise to all the mesodermal structures of the embryo. The clear hyaloplasm is destined to become ectoderm and the remaining yolk-filled cytoplasm is presumptive endoderm. The precleavage egg is, therefore, oriented along anatomical lines, and its cytoplasm is already segregated into three distinct presumptive areas, each of which will

Figure 7-4. Formation of the mesodermal crescent. As the entering sperm nucleus moves towards the female pronucleus, it drags along with it some of the cortical cytoplasm. The cortical cytoplasm remaining on the egg surface moves around the egg until it forms a crescent of material near the original sperm entrance point. Illustration D is an external view of the egg after the formation of the mesodermal crescent.

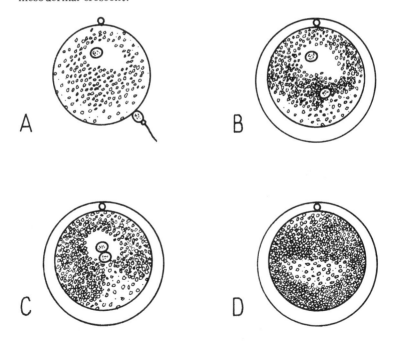

develop normally into one of the three definitive germ layers of the adult. This precleavage egg with its localized parts can be seen in Figure 7-5.

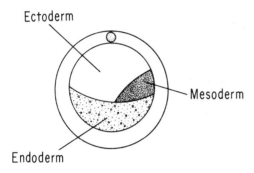

Figure 7-5. Fate map of an uncleaved egg showing the localization of the potential germ layers. The white area is future ectoderm, the closely stippled area is potential mesoderm, and the widely stippled region will develop into endoderm.

Cleavage Pattern

The first cleavage plane (Fig. 7-6B, C), which occurs within an hour and a half after fertilization, is oriented from animal to vegetal pole. It also cuts through the sperm entrance point which is now represented by the mesodermal crescent. The first cleavage, therefore, occurs along the already established median plane of the embryo and separates it into two lateral halves. The result of this cleavage is two identical blastomeres.

The second cleavage plane is also oriented from the animal to vegetal pole, at right angles to the first. It is, however, displaced slightly posterior to the animal-vegetal axis (Figs. 7-6D, 7-6E). The 4-cell stage, therefore, is composed of two larger anterior and two smaller posterior blastomeres. The mesodermal crescent is located in both posterior blastomeres. The four blastomeres are not in intimate contact throughout their entire surface. A cavity, the beginning of the blastocoele, is found enclosed by their mesial surfaces (Fig. 7-6E).

The third cleavage is horizontal, slightly displaced toward the animal pole. This creates four animal micromeres and also four vegetal macromeres. The two anterior micromeres and macromeres are larger than the two posterior micromeres and macromeres, respectively. This is due to the inequality produced by the second cleavage plane. The mesodermal crescent is now located in the two posterior macromeres (Figs. 7-6F, G).

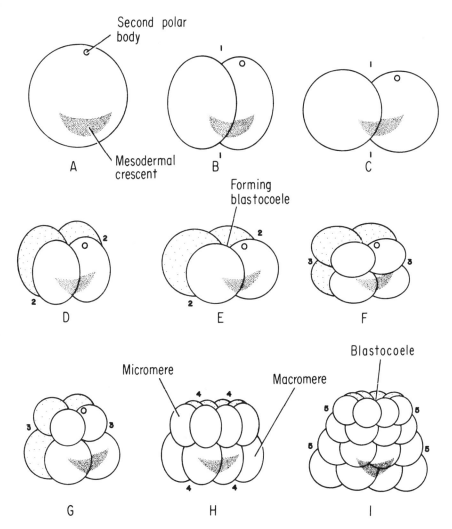

Figure 7-6. Cleavage planes of the Amphioxus egg. (A) The uncleaved egg, showing the position of the mesodermal crescent and the location of the animal pole as indicated by the location of the second polar body. (B-C) The first cleavage plane is oriented from animal to vegetal pole. It cuts through the mesodermal crescent and separates the original egg into two identical blastomeres. (D-E) The second cleavage plane is also oriented from animal to vegetal pole but at right angles to the first. This plane, however, is displaced slightly posterior to the animal-vegetal axis. The 4-cell stage, therefore, is composed of two larger anterior and two smaller posterior blastomeres. (F-G) The third cleavage is a horizontal one, which is slightly displaced toward the animal pole. The result is four animal micromeres and four vegetal macromeres. (H) Each of the existing blastomeres is further subdivided by the fourth cleavage plane. The result is an animal hemisphere of eight micromeres and a vegetal hemisphere of eight macromeres. (I) The 32-cell stage. Each of the micromeres and macromeres is divided latitudinally by a cleavage plane. The central cavity, or blastocoele, can be seen.

The fourth cleavage is a double plane—each one oriented the same as planes one and two, i.e., from animal to vegetal pole (Fig. 7-6H). This results in eight animal micromeres and eight vegetal macromeres. These cleavage planes, however, do not divide each of the blastomeres equally. The most anterior and posterior blastomeres of both the animal and vegetal hemispheres are slightly smaller than the others.

The fifth cleavage is latitudinal, i.e., parallel to plane three. It also is a double plane, one occurring in the animal, the other in the vegetal hemisphere. This has the effect of dividing the micromeres and macromeres into two rows of eight cells each. This 32-cell stage can be seen in Figure 7-6I. The cleavage pattern beyond this is irregular and difficult to follow. The vegetal cells do, however, because of their yolk content, reproduce at a slower rate than those in the animal hemisphere. This results in a definite blastomere size difference, those in the vegetal hemisphere being larger than those toward the animal pole.

Blastula Formation and Egg Orientation

During cleavage, the central cavity between the blastomeres continues to enlarge. This cavity at first is open to the exterior of the egg between all the blastomeres. As cleavage continues, the blastomeres crowd together, closing up most of these openings. Two, however, remain longer than the rest. One is at the animal pole, and one at the vegetal pole (Fig. 7-6G). The opening at the vegetal pole can be seen to be larger than the one at the animal pole. Soon, both these openings also close. Once these openings close, the egg enlarges approximately one-third its fertilization size. This "growth" is due primarily to an increase in the size of the blastocoele.

At the blastula stage the egg is oriented as in Figure 7-7. The polar body still marks the animal pole of the egg. Consequently, an animal-vegetal axis can be drawn. This is represented in Figure 7-7 by a dotted line. The cells on the side of the blastula contain the mesodermal crescent. The anterior-posterior axis is represented in Figure 7-7 by the solid line A-P.

A sagittal section through a blastula of this stage is shown in Figure 7-8. The large cells anterior to the mesodermal cells on the ventral surface are future endodermal cells, whereas those lying anterior to the crescent on the dorsal surface are presumptive ectoderm. The blastopore will occur at approximately the junction

Figure 7-7. The blastula of the Amphioxus egg. The dotted line represents the original animal-vegetal axis; the solid line represents the future anterior-posterior axis. Note the location of the mesodermal crescent (stippled area) in the posterior region of the blastula.

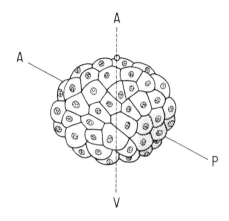

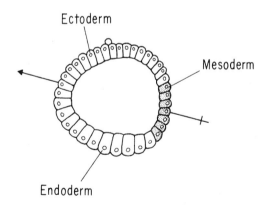

Figure 7-8. Sagittal section of the Amphioxus blastula. This should be compared with Figure 7-7. The arrow indicates the anterior-posterior axis. Note the position of the polar body denoting the position of the original animal pole. The stippled areas represent the future mesodermal cells. Anterior and dorsal to these are the ectodermal cells, whereas ventral and posterior are the presumptive endodermal cells.

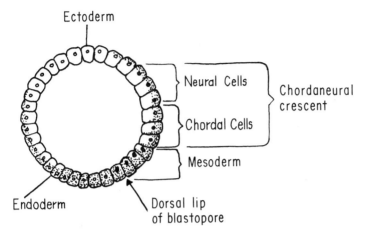

Figure 7-9. Chordaneural crescent. Those cells lying just anterior and dorsal to the mesodermal crescent will form notochord, and those lying dorsal to these will form the neural plate. Collectively, these latter two cell types constitute a chordaneural crescent.

between the endodermal and mesodermal cells. Those cells lying just anterior to the mesodermal crescent on the dorsal surface will form the notochord, and those lying dorsal to these will form the neural plate. Collectively, these latter two cell types constitute a chordaneural crescent (See Figure 7-9).

Gastrulation

The onset of gastrulation is marked by a thickening of the endoderm ventral to the mesodermal crescent. This endodermal plate is composed of the largest cells of the embryo, cells which were derived from the original vegetal pole region. It is this plate of endodermal cells that begins to invaginate. The ectodermal and mesodermal cells continue to reproduce at a rapid rate. Figure 7-10 shows the appearance of the early gastrula during the formation of the blastopore. Invagination continues until the endoderm comes into intimate contact with the overlying ectoderm. The blastocoele, therefore, is nearly obliterated. At the completion of the gastrulation process, the anterior-posterior axis passes through the blastopore itself. Once this stage of embryonic development has been reached, the developing organism is referred to as a larva.

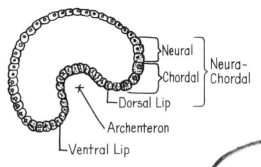

Figure 7-10. An early Amphioxus gastrula showing the newly formed archenteron and blastopore.

Ectodermal Development

In the early gastrula, six rows of cells immediately anterior to the dorsal lip of the blastopore are important for notochord and neural tube development. The first three rows are chordal cells, whereas the three next most anterior rows are destined to become the neural plate (Figs. 7-9 and 7-10). The chorda cells invaginate around the edge of the dorsal lip of the blastopore, pushing inward in such a manner that they split the already invaginated mesoderm. The notochord, therefore, becomes the middorsal wall of the archenteron. The neural plate cells remain on the outer surface. As the larva grows in length, both the notochord and neural plate cells reproduce rapidly, extending anteriorly from the dorsal lip of the blastopore along the middorsal line (Fig. 7-11). Grooves soon appear along the roof of the archenteron. These grooves deepen, allowing the development of both notochord and mesodermal pouches. This entire

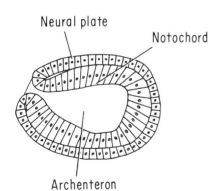

Figure 7-11. Sagittal view through a late gastrula. Note that the notochordal cells have invaginated around the dorsal lip of the blastopore. Both these cells and the neural plate cells remaining on the surface continue to reproduce rapidly as the embryo grows in length. The notochord comes to lie beneath the neural plate.

formation can be seen in Figure 7-12. As this notochordal development proceeds, the individual cells begin to take on a unique appearance, filling with a clear "cartilagelike" material. This notochord is positioned directly beneath the neural plate cells on the surface of the larva.

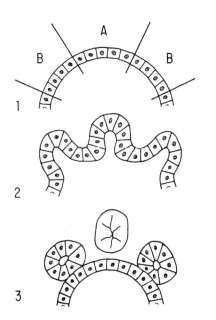

Figure 7-12. Diagrammatic representation of the development of the notochord and somites from the dorsal roof of the archenteron. (1) The roof of the archenteron can be divided into three distinct sections, indicated in the drawing by A and B. (2) As development proceeds, both sections A and B develop grooves within them. (3) This continues until each separates from the archenteron. The result is a dorsally located notochord and two laterally located somites. The roof of the archenteron again closes over, forming the definitive gut.

The ectoderm beneath the ventral lip now begins to grow dorsally, covering the blastopore completely. This ectodermal growth continues anteriorly, covering the already formed neural plate. The archenteron, however, continues to communicate with the outside via the newly formed space between the overgrowing ectodermal sheet and the medullary plate. This anterior opening, which advances anteriorly as growth continues, is termed the anterior neuropore (Fig. 7-13). After the medullary plate is covered, the edges of the plate begin to round up and approach one another in the midline, forming the neural tube (Fig. 7-14). This neural tube contains a cavity, the neural canal or neurocoele, which is open anteriorly at the anterior neuropore and is continuous posteriorly with the original archenteron, through the neurenteric canal (Fig. 7-15). Both the anterior neuropore and the neurenteric canal remain open throughout most of the embryonic period.

Figure 7-13. Formation of the anterior neuropore. As the sheet of ectodermal cells proceeds anteriorly from the ventral lip of the blastopore, it creates an opening between it and the underlying neural plate. This opening is the anterior neuropore.

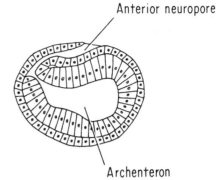

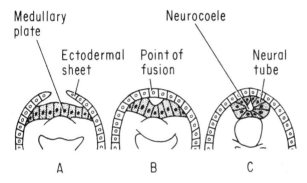

Figure 7-14. Formation of the neural tube. (A) Each of the anteriorly growing ectodermal sheets begins to approach the other in the dorsal midline of the embryo. (B) These sheets finally fuse, forming a complete covering over the medullary plate. This medullary plate itself then begins to round up at the edges, forming a completed neural tube. (C) The enclosed canal is the neurocoele. This is continuous with the archenteron (See Fig. 7-15).

Figure 7-15. The neurenteric canal. The cavity of the neurocoele is continuous with the archenteron in the region of the original blastopore. This connection between the neurocoele and the archenteron is referred to as the neurenteric canal.

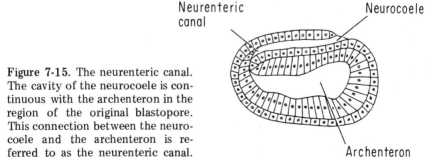

Mesodermal Development

During the gastrulation process, the cells of the mesodermal crescent invaginate as part of the roof of the archenteron. They are located in the gastrula in the dorsolateral roof of the archenteron, bordered dorsally by the notochordal cells and ventrally by the large endodermal cells. A groove begins to develop within these mesodermal ridges which gradually deepens and finally pinches off as two rows of mesodermal tubes (Fig. 7-12). These tubes then segment to give rise to the somites. This segmentation of the tube begins anteriorly and proceeds posteriorly. The cavity formed within the somite is the myocoele. This originally was continuous with the gut cavity. The segmentation of the tubes into somites does not occur with bilateral symmetry. The left somites alternate with those of the right, the left ones being ahead (Fig. 7-16). At the 15-somite stage, the original somite material begins to differentiate. The first mesoderm to form is the lateral mesoderm. This forms at the ventral edge

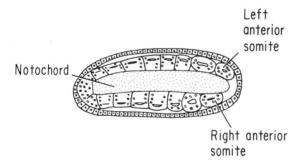

Figure 7-16. Frontal section through the neurula. As the embryo increases in length, the somites begin to segment. This segmentation occurs first in the anterior end of the animal and proceeds progressively posteriorly. It should be noted that the segmentation is not bilaterally symmetrical. The left somites alternate with those on the right.

of the original somite and grows ventrally around the gut cavity in two layers. The outer layer, that closest to the ectoderm, is the somatic mesoderm; the inner layer, that closest to the endoderm, is the splanchnic mesoderm. The cavity between the two layers of mesoderm is the splanchnocoele. This will be the coelomic cavity of the adult. Each layer continues to grow ventrally until it meets the corresponding layer from the other side. When this fusion takes place, a completed coelomic cavity is formed. The combination of endoderm and splanchnic mesoderm is referred to as the splanchnopleure, whereas the term "somatopleure" is reserved for the ectoderm and somatic mesoderm. The somite itself undergoes a period of differentiation. The original myocoele separates the somite into an outer portion and an inner one. The outer portion is the dermatome. This eventually grows ventrally to form the dermis of the skin, whereas the inner layer, the myotome, differentiates into muscle cells, which ultimately give rise to the musculature of the body. From the base of the myotome another group of cells pinches off and grows dorsally between the myotome and the notochord and neural tube. This sclerotome will develop into the skeletogenous sheath which surrounds both the notochord and neural tube. These changes can be seen in Figure 7-17.

Figure 7-17. Development of the somite. The first mesoderm to form is the lateral mesoderm. This forms at the ventral edge of the original somite and grows ventrally around the gut cavity in two layers. The outer layer is the somatic mesoderm. The inner layer is the splanchnic mesoderm. The cavity between the two layers of mesoderm is the splanchnocoele or coelome. Each layer continues to grow ventrally until it meets the corresponding layer from the other side. The combination of endoderm and splanchnic mesoderm is referred to as the splanchnopleure, whereas the term somatopleure is reserved for the ectoderm and somatic mesoderm. The myocoele separates the somite into an outer dermatome and an inner myotome. From the base of the myotome, another group of cells pinches off and grows dorsally between the myotome and the notochord and neural tube. This is the sclerotome. It develops into the skeletogenous sheath, which surrounds both the notochord and neural tube.

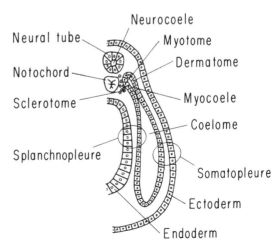

Development of the Endoderm

After the somites have formed, the dorsal portion of the anterior end of the gut cavity forms a new diverticulum (Fig. 7-18A). This original single pouch soon splits into a separate left and right diverticulum (Fig. 7-18B). These are called the left and right dorsal diverticula. Each is then cut off from the gut cavity itself. The left diverticulum remains rather small. Its wall stays relatively thick. An ectodermal invagination soon grows into this thick-walled left dorsal diverticulum to form an adult sense organ, the preoral pit (Fig. 7-19).

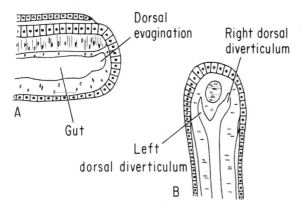

Figure 7-18. Formation of the left and right dorsal diverticula. (A) At the anterior end of the original gut, an evagination occurs. (B) This is a frontal section of the same embryo represented in A. The original dorsal pouch soon splits into separate left and right dorsal diverticula.

Figure 7-19. Formation of the preoral pit. (A) The left dorsal diverticulum is separated from the gut cavity. Its walls stay relatively thick. (B) An ectodermal invagination soon grows into this thick-walled left dorsal diverticulum to form the preoral pit.

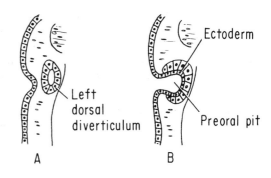

The right dorsal diverticulum becomes very thin-walled and grows to occupy eventually nearly all the space in the anterior end of the embryo between the ectoderm and notochord. In reality, therefore, the right dorsal diverticulum becomes the head cavity.

Once the left and right dorsal diverticula separate from the anterior end of the gut, an evagination forms in the ventral anterior area. This extension of the ventral end of the gut cavity differentiates into the endostyle of the adult (Fig. 7-20). Still another evagination on the right side of the gut, slightly posterior to the endostylar evagination, develops. This evagination grows both ventral and dorsal to form the club-shaped gland (Fig. 7-20). The club-shaped gland has no known function, although it has been postulated that its glandular cells secrete a substance that aids in separating food from sand, both of which are ingested at the same time.

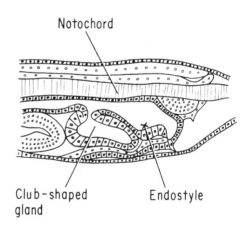

Figure 7-20. Anterior end of Amphioxus, showing the position of the club-shaped gland relative to the notochord and endostyle.

The lumen of the gut posterior to the evagination of the club-shaped gland is not circular but oblong. Its dorsal-ventral dimension is much greater than its lateral. In this area the branchial apparatus, the forerunner of the gills, begins to develop as a thickening in the wall and floor of the gut (Fig. 7-21). Invaginations of the ectoderm and evaginations of the endoderm in the area of this branchial apparatus form the gill slits. These gill slits open directly to the outside. On the left side of the posterior region of the gut, another dual invagination-evagination of ectoderm and endoderm occurs, forming the anus (Fig. 7-22). The mouth is formed in a similar fashion at the extreme anterior end of the gut. Both the mouth and anus are found a little to the left of the midventral line.

84 The Development of Amphioxus

Figure 7-21. Section through the pharyngeal region showing the developing branchial apparatus. The gill slit is formed by a concomitant evagination of the pharyngeal endoderm and an invagination of the ectoderm.

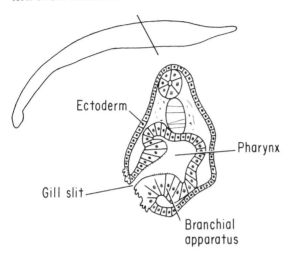

Figure 7-22. Formation of the anus. (A) An evagination from the posterior end of the gut meets the proctodeal invagination of the ectoderm. (B) This soon breaks through forming a continuous anal opening between the posterior end of the gut and the exterior of the animal.

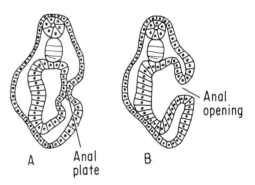

There is no reason in a book of this scope to go further than this with the development of Amphioxus. The basic organ systems have already been established, and subsequent development would not enhance our understanding of vertebrate development.

"Thy word is a lamp unto my feet, and a light unto my path. The entrance of Thy words giveth light; it giveth understanding unto the simple." Psalms 119:105, 130.

Chapter **8**

The Embryology of Teleosts

The fishes constitute such a large segment of the phylum Chordata that it would be impossible, in a single chapter of a book devoted to the "elements" of vertebrate embryology, to discuss the development of this entire group. Four of the eight vertebrate classes are "fishes." These range from the jawless Agnathan forms to the highly specialized Osteichthyian types. Even within the order Teleostiae of the class Osteichthyes (most modern fish), we find great variation in egg morphology and, consequently, of the developmental pattern. Serious students of embryology can find numerous references to the embryological development of the most popular aquatic forms.

A modern fish that has attracted much attention is the Fathead minnow (*Pimephales promelas*). It is important for it is extensively used as an assay fish for aquatic pollution studies. Its early embryology has recently been described.* For the purpose of this chapter, only a generalized description of Teleost development will be attempted. It is hoped that by using this basic approach, the embryonic development of any specific fish will be more readily understood.

---

*Harold W. Manner and Casimira M. Dewese, 1974, Early Embryology of the Fathead Minnow *Pimephales promelas* Rafinesque, Anat. Rec. 180:99-110.

Egg Morphology

Most teleost eggs have a size exceeding 0.7 mm. Their large size is due primarily to the exceptionally large quantity of yolk in the unfertilized egg. These eggs would, therefore, be classed as polylecithal on the basis of yolk content. There are, however, egg variations among the various teleost species, variations not only in the amount of yolk but also in the relationship between the yolk (deutoplasm) and the developmental cytoplasm. As a consequence of the great amount of yolk, cleavage is always partial (meroblastic or discoidal). Embryonic development, therefore, is limited to the cytoplasm that accumulates near the animal pole of the egg. The distribution of the yolk is not that of the typical polylecithal egg (e.g., the chick) for, in the teleosts, it is concentrated toward the center of the egg rather than at one end. Eggs of this type are referred to as centrolecithal. During the prefertilization stages, the cytoplasm around the edge of the yolk forms a peripheral layer. This is fairly constant, except in the region of the animal pole where a heavier concentration of yolk-free cytoplasm can be seen. This large area of yolk-free cytoplasm is associated with the presence of the female pronucleus. During the development of the egg in the female, an opening in the chorion or "egg shell" near the animal pole is formed. This opening, the micropyle, is structurally designed to facilitate sperm entrance. The area of yolk-free cytoplasm is referred to as the gel layer. It is extremely sticky on its inner surface. This surface gel layer is covered by the vitelline membrane. The vitelline membrane itself is highly specialized, being composed of two and sometimes three distinct layers. A typical teleost egg, prior to fertilization, but after spawning, can be seen in Figure 8-1.

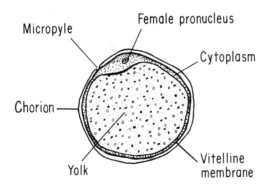

Figure 8-1. A typical teleost egg, prior to fertilization, but after spawning.

Fertilization

Although examples of both external and internal fertilization among the teleosts can be cited, the majority exhibit external fertilization. The water into which the egg is deposited initiates a surface reaction. After fertilization, the vitelline membrane lifts from the surface of the egg. This is accompanied by a compensatory shrinking of the egg itself. This shrinking creates a space between the egg surface and the vitelline membrane (Fig. 8-2). This perivitelline space is filled with a fluid produced by the egg itself. The egg is capable of rotating freely, once the perivitelline space is created. In most species, the rotation of the egg results in an animal pole which is uppermost. This is due primarily to the weight of the yolk in the vegetal hemisphere. The sperm, which have been deposited near the egg, soon surround it. One enters through the already formed micropyle. Fertilization in the teleosts is, therefore, of the monospermy variety. Although fertilization is considered by most authorities to be the union of the male and female pronuclei, the processes or events of fertilization actually begin as soon as the sperm enters the micropyle. The egg is still not completely mature at the time of sperm entrance, being in the latter stages of the second meiotic division. This meiotic process is soon completed, and the resulting polar body is extruded to the surface. The nuclear materials of the egg then reform into the female pronucleus. When in this state, the male pronucleus is now able to unite with the female pronucleus, and the act of fertilization is completed.

A change can also be seen, during the fertilization process, in the relative distribution of both the deutoplasmic and cytoplasmic

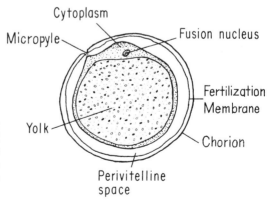

Figure 8-2. A teleost egg after fertilization. The shrinking of the egg, together with the lifting of the vitelline membrane, creates the perivitelline space.

components of the egg. The entrance of the sperm causes a reaction within the surface cytoplasm. Whereas this was formerly distributed rather ubiquitously over the surface of the egg, it now has a tendency to flow toward the point of sperm entrance. This establishes a protoplasmic disk (germinal disk) at the animal pole and reduces the cytoplasmic covering over the rest of the egg to a fine gel layer. In terms of egg classification based on yolk distribution, the egg, during the fertilization process, is converted from a centrolecithal to a heavily telolecithal type.

In addition to initiating the developmental processes, fertilization restores the diploid number of chromosomes. By so doing, the paternal genetic traits, carried on the chromosomes of the sperm nucleus (male pronucleus), and the maternal genetic traits, carried on the chromosomes in the egg nucleus (female pronucleus), are united in the fusion nucleus of the fertilized egg.

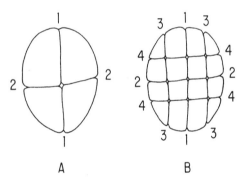

Figure 8-3. Cleavage planes in the teleost egg. Only the germinal disk cleaves. (A) The first and second planes. (B) The disk at the completion of the first four cleavages.

Cleavage and Blastula Formation

Due to the heavily telolecithal yolk distribution, cleavage in the teleost egg is meroblastic or discoidal. Only the cytoplasmic (germinal) disk, located at the animal pole, undergoes cleavage. The first six cleavage planes are fairly regular in all teleost eggs. These planes and their orientation can be seen in Figure 8-3. Between thirty minutes and one hour after fertilization, the first cleavage plane can be seen. This cuts the germinal disk into two equal blastomeres. The second cleavage plane occurs at right angles to the first, separating the germinal disk into four blastomeres of approximately equal size. The third cleavage planes are oriented parallel to plane one and at right angles to plane two. The result is eight blastomeres of fairly equal size. The

Figure 8-4. Cross section through the blastula. The periblast moves out over the surface of the yolk and eventually completely surrounds it, forming a yolk sac.

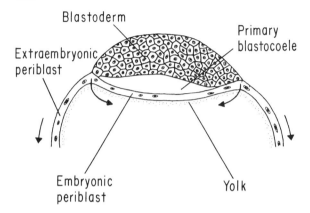

fourth cleavage planes are oriented parallel to cleavage plane two and at right angles to planes one and three. This results in the 16-cell stage, composed of four rows of four blastomeres each. The fifth plane is actually four separate planes. Each one occurs parallel to planes one and three and at right angles to planes two and four. This 32-cell stage is very typical of the teleost egg. The sixth cleavage plane is horizontal, i.e., it begins to separate the original single-layered blastoderm into two layers. From this point on, the cleavage planes become fairly irregular. They do continue, however, until a mound of cells is built up at the animal pole. This stage, represented by this mound of cells located at the animal pole, is properly called the blastula. The mound of cells is the blastoderm. Cells from the margin of the blastoderm begin to migrate both medially and away from the blastoderm. During this migration, the cells lose their membranes and the nuclei continue to move throughout the gel layer surrounding the yolk. This new multinucleated layer is a syncytium and in the Teleost is known as the periblast. That portion which is beneath the blastoderm is the embryonic periblast. The remainder, that surrounding the yolk, is the extraembryonic periblast (See Fig. 8-4). Eventually, this periblast completely surrounds the yolk forming a yolk sac. The cavity that forms between the embryonic periblast and the blastoderm proper is the segmentation cavity, or blastocoele.

Gastrulation

Fate Map of the Early Gastrula

Although the individual cells of the blastoderm are at this stage pluripotential, that is, capable of developing into many different tissues, depending on the morphogenetic stimuli to which they are subjected, they have been followed, by means of vital stains, to a definite position in the adult fish's body. Maps made utilizing this technique are referred to as fate maps. In essence, they indicate what the cells of the blastoderm will become (their prospective fate) if left to the normal embryological processes. A fate map of a typical early teleost gastrula is seen in Figure 8-5. One margin of the blastoderm is thickened slightly. This area represents the future posterior end of the embryo. From this point anteriorly, in the median line of the blastoderm, a series of presumptive tissues can be seen. Most posteriorly is the presumptive endoderm. Immediately anterior to this is the presumptive prechordal material. This is followed by the future notochordal material. Just anterior to the presumptive notochord is the neural ectoderm. This will ultimately become the neural tube of the central nervous system. Surrounding these presumptive, organ-forming areas of the midline are vast expanses of two different

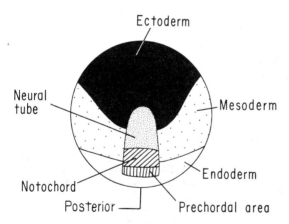

Figure 8-5. Fate map of a teleost blastula depicting the presumptive areas. These maps are made by applying vital stains at this stage and watching their movements during subsequent development.

presumptive tissue materials. Anterior to the neural ectoderm and appearing as a semicircular mass is the future ectoderm of the skin. Posterior to this ectoderm, and located on each side of the neural ectoderm and notochord, is the presumptive mesoderm. By recognizing the position of each of these presumptive tissue areas, the gastrulation movements and the ultimate distribution of these tissues in the postgastrular embryo should be more readily comprehended.

Gastrulation Movements

As gastrulation begins, the entire peripheral rim of the blastoderm begins to thicken. This thickening is called the germ ring. It is more pronounced in the posterior region, the site of the future dorsal lip of the blastopore, than it is in the rest of the germ ring. This characteristic posterior thickening of the germ ring is referred to as the embryonic shield in teleosts. The other regions of the germ ring might also be considered as lips of the blastopore. An epibolic movement causes the cells of the blastoderm to move around the yolk over the surface of the periblast. Figure 8-6 shows the direction of this movement. In the region of the dorsal lip, however, the ventral progression of the cells is interrupted, and the presumptive organ-forming areas, which were delineated in the fate map, begin to involute around the dorsal lip. The first cells to move into the interior are the endodermal cells. These involute around the dorsal lip of the blastopore and begin to grow inward over the surface of the periblast. This new endodermal layer forms the hypoblast of the gastrula. The original blastoderm can now be called the epiblast.

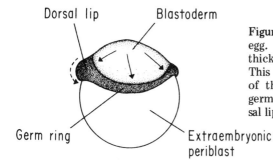

Figure 8-6. Gastrulation in a teleost egg. The rim of the blastoderm thickens to become the germ ring. This moves down over the surface of the periblast. At one point the germ ring thickens. This is the dorsal lip or the embryonic shield.

The prechordal tissue is the next presumptive area to move around the dorsal lip, followed immediately by the presumptive notochord. As they move in, they grow between the hypoblast and the overlying epiblast. They become compressed and eventually form a thin cord of cells in the dorsal midline of the developing embryo. The mesoderm, in a similar fashion, involutes around the dorsal, lateral, and ventral lips and becomes inserted between the overlying epiblast and the underlying hypoblast. These two layers, i.e., the notochord and the mesoderm, form a continuous sheet of cells. Figure 8-7 represents a cross section through a late gastrula, showing the distribution of these layers. The notochord, however, soon becomes isolated from the mesodermal areas. The mesoderm, as it involutes, grows toward the midline. This results in a thickened mesodermal mass on each side of the notochord. These thickened

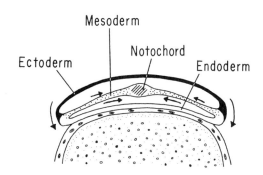

Figure 8-7. Cross section through a teleost gastrula. The involuting germ ring forms the endoderm, mesoderm, and notochord.

masses soon become the somites, whereas the remainder of the mesoderm is now known as the lateral plate mesoderm. While these processes of involution are occurring, the rest of the germ ring continues to grow down and around the yolk as a result of the continuing epibolic process. These germ ring areas soon completely enclose the yolk when they meet in the vicinity of the dorsal lip (Fig. 8-8). The presumptive neural tube area also continues to thicken so that, at the end of gastrulation, the embryo's outline can be fairly distinctly seen when viewed dorsally (See Figure 8-9). A sagittal section through the same embryo is shown in Figure 8-10.

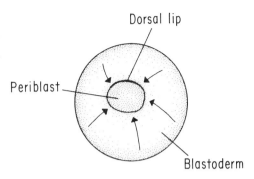

Figure 8-8. The germ ring of the blastoderm continues to grow in an epibolic manner around the egg. The edges of the germ ring meet in the vicinity of the original dorsal lip.

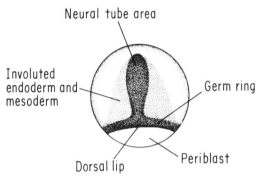

Figure 8-9. Origin of the embryonic axis. The presumptive neural tube area continues to thicken, so that at the end of gastrulation the embryo's outline can be fairly distinctly seen.

Figure 8-10. (A) Transverse section through an embryo of the stage depicted in Figure 8-9. All the primary tissues can be seen. (B) The same embryo, as seen in a midsagittal section.

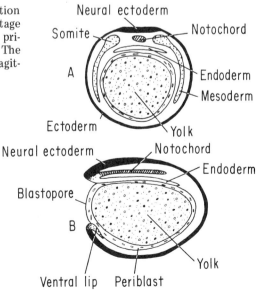

Neurulation

The process of neurulation in the teleosts is different from that found in most other vertebrates. The neural plate found on the surface of the gastrula in the dorsal midline gradually narrows. This narrowing is due to a sinking of some of the cells from the surface. This continues until the presumptive neural tube becomes a thickened cord of cells lying in the anterior-posterior axis of the body, immediately above the notochord. Eventually, the sinking neural tube is overgrown with epidermis. The lumen of the neural tube develops only secondarily. The entire process of neurulation in teleosts is graphically depicted in Figure 8-11. A sagittal section through a teleost embryo after the formation of the neural tube can be seen in Figure 8-12.

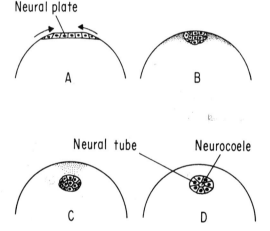

Figure 8-11. Process of neurulation in the teleost embryo. (A) The lateral cells of the neural plate push toward the midline. This continues until the presumptive neural tube becomes a thickened cord of cells lying in the anterior-posterior axis of the body. (B) Eventually this cord of cells sinks from the surface (C) and is overgrown with epidermis. The solid neural cord then develops a lumen, the neurocoele (D).

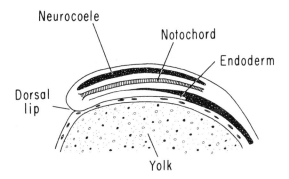

Figure 8-12. Sagittal section through a teleost embryo after the formation of the neural tube. The stippled black areas represent the original blastocoele and the neurocoele.

Formation of the Primitive Gut

The hypoblast (after gastrulation this should now be called the endoderm), which now lies as a narrow sheet of cells immediately above the periblast, begins to fold upward in the median ventral line of the embryo to form the beginning of the gut tube (Fig. 8-13). This upward folding of the endoderm continues until a hollow tube

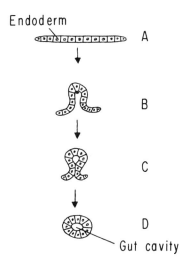

Figure 8-13. Diagrammatic representation of the formation of the primitive gut cavity. After gastrulation, the endoderm lies as a narrow sheet of cells in the midline above the periblast. (A) This sheet of cells then folds upward to form a hollow tube running from a position near the anterior edge of the blastoderm to the dorsal lip (B, C, D).

running from a position near the anterior edge of the blastoderm to the dorsal lip is formed. This primitive tube is the primary gut (enteron) of the developing teleost. This process of gut formation occurs concomitantly with the development of the neural tube. A sagittal section through an early postneurula embryo, showing the completed neural and endodermal tubes, is depicted in Figure 8-14.

Figure 8-14. Midsagittal section through an early postneurula teleost embryo, showing the relative positions of the newly formed neural tube and gut. The stippled area represents cavities within the neural tube and primitive gut.

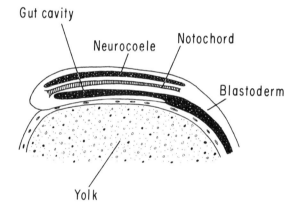

Early Mesodermal Differentiation

The lateral mesoderm, which immediately after gastrulation was found lateral to the somites between the endoderm and ectoderm (the name now applied to the epiblast), begins to delaminate, i.e., separate into an inner and an outer layer. The inner layer is the splanchnic mesoderm, whereas the outer layer is the somatic mesoderm. The space between them is the coelomic cavity. Both the left and right halves of the somatic and splanchnic mesoderm continue to grow ventrad around the periblast-lined yolk. Eventually, each meets its corresponding layer of the other side to which it fuses. The yolk is, therefore, lined with periblast, splanchnic mesoderm, somatic mesoderm, and ectoderm after these mesodermal developmental processes take place. These changes are reflected in Figure 8-15 which represents a typical cross section through an entire postneurula teleost embryo.

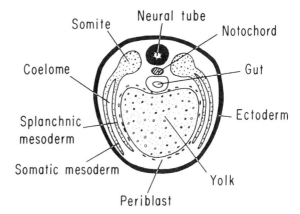

Figure 8-15. Cross section through a late post-neurula embryo showing the relationships of the tissues to one another after mesoderm delamination.

Formation of the Body Shape

The embryo at this stage is still confined within the egg shell. Consequently, any changes in the body contour, before hatching, are limited by the closely applied chorion. From the original region of the dorsal lip of the blastopore, a growth, the tail bud, begins to form. This formation continues to grow posteriorly between the ectoderm of the yolk and the overlying shell. The neural tube, notochord, mesoderm, and gut also grow by cellular proliferation for some distance within the caudal or tail fold. An embryo undergoing this tail development is illustrated in Figure 8-16. The metabolic

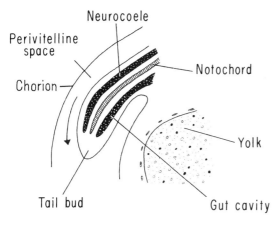

Figure 8-16. The tail bud, which begins to form from the original region of the dorsal lip of the blastopore, continues to grow posteriorly between the ectoderm of the yolk sac and the overlying chorion. The neural tube, notochord, mesoderm, and gut grow by cellular proliferation for some distance within this caudal or tail fold.

98   The Embryology of Teleosts

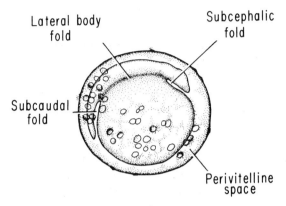

Figure 8-17. Lateral view of the entire embryo prior to hatching. The decrease in the size of the yolk sac, together with the growth of the head and tail buds, gives definite form to the prehatching embryo.

utilization of the yolk and the consequent decrease in the amount remaining necessitate a contraction of those tissues comprising the yolk sac. This decrease in size, together with an undercutting by lateral and subcephalic folds, gives a definitive body shape to the embryo. A lateral view of the embryo after the establishment of these folds is given in Figure 8-17. The yolk sac, however, continues to be an extremely obvious structure in the early embryo and larvae. However, as the yolk continues to be utilized metabolically by the developing teleost, the yolk sac wall recedes until finally it reaches a level continuous with that of the rest of the body (Fig. 8-18).

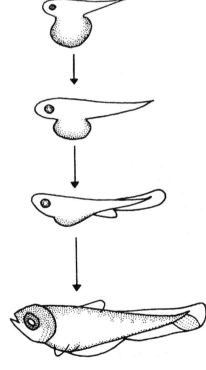

Figure 8-18. As the yolk continues to be utilized metabolically by the developing teleost, the yolk sac wall recedes until it reaches a level continuous with that of the rest of the body.

Early Development of the Neural Tube

Concomitant with the increase in body length is the early differentiation of both the neural tube and primitive gut. Anteriorly, in the head region, the neural tube enlarges somewhat to indicate the presence of the future brain. This original brain enlargement undergoes a further subdivision into the three primary brain regions: the anterior prosencephalon, middle mesencephalon, and posterior rhombencephalon. The eye, a very conspicuous structure in the young teleost, develops as an evagination of the lateral wall of the posterior portion of the prosencephalon. This early development of the anterior end of the neural tube can be seen in Figure 8-19. In the trunk and tail region, the neural tube remains narrow. This portion will ultimately become the spinal cord.

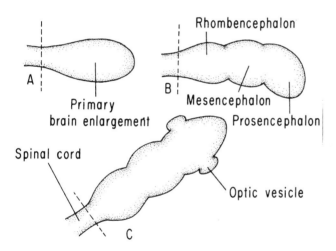

Figure 8-19. Early development of the teleost brain. (A) The anterior end of the neural tube enlarges to form the primary brain enlargement. (B) This soon subdivides into an anterior prosencephalon, a middle mesencephalon, and a posterior rhombencephalon. (C) The eye develops as an evagination of the lateral wall of the posterior part of the prosencephalon.

Early Development of the Primitive Gut

The primitive gut, after elongation of the anterior-posterior body axis, can now be described as a three-part digestive tube. That portion anterior to the yolk sac region is the foregut. The caudal part, i.e., the portion of the tube posterior to the yolk sac, is the hindgut. The remainder, between the foregut and the hindgut, is the midgut.

The blind anterior end of the foregut soon develops two evaginations, one anterior-dorsal, the other anterior-ventral (Fig. 8-20). The anterior-dorsal diverticulum is the preoral gut. This is, however, only a temporary structure in vertebrate development. The anterior-ventral evagination meets an ectodermal invagination (the stomadeum) which is developing in the ventral head region. These two processes soon meet, forming the oral plate. Shortly after, the oral plate ruptures, producing the mouth opening. It should be noted here that, as a result of this mechanism of formation, the mouth or oral cavity is lined with both ectoderm and endoderm. This process of oral cavity and mouth formation is graphically depicted in Figure 8-20.

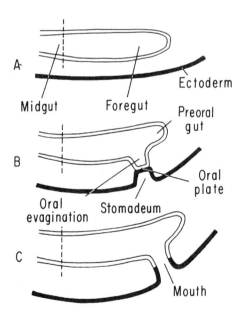

Figure 8-20. Development of the teleost mouth. (A) The foregut is the site of mouth development. (B) The blind anterior end of the foregut develops antero-dorsal (preoral gut) and antero-ventral (oral) evaginations. The oral evagination meets the forming stomadeum to produce the oral plate. (C) The oral plate then ruptures to form the mouth opening. As a result of this method of formation, the mouth or oral cavity is lined with both ectoderm and endoderm.

In the posterior region of the hindgut, two evaginations, similar in structure and location to those in the foregut, make their appearance. One is posterior-dorsal, one posterior-ventral. The posterior-dorsal diverticulum is a transitory structure, the postanal gut. This will ultimately degenerate. The posterior-ventral evagination, however, grows toward an ectodermal invagination (the proctodeum), which is located posterior to the yolk sac (Fig. 8-21). The posterior-ventral evagination of the hindgut meets and fuses with the proctodeum, forming the anal or cloacal plate. This then ruptures, completing the cloacal opening. Once this occurs, the digestive tube is complete, with an anterior mouth opening, and a posterior cloacal orifice. This posterior early development of the hindgut is pictured in Figure 8-21.

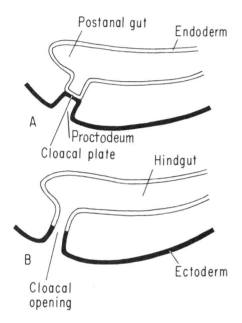

Figure 8-21. Formation of the cloaca in the teleost. (A) The endodermal cloacal evagination meets the ectodermal invagination (the proctodeum) to produce the cloacal or anal plate. (B) When this plate ruptures, the cloacal opening is produced.

Early Circulatory Pattern

As the yolk, enclosed in the yolk sac, is not directly connected to the primitive gut, an early circulatory pattern must be established to transport this nutritive material to all parts of the embryonic body. This early circulatory pattern begins with two omphalomesenteric veins which originate in the yolk sac. Beneath the foregut they fuse, forming the pulsating heart. The subintestinal arteries leaving the heart, course ventral to the foregut, until they reach its most anterior end. They then proceed dorsally around the anterior end of the preoral gut. These vessels, now in a dorsal position, are referred to as the supraintestinal arteries, the forerunners of the adult dorsal aortae. In the posterior region, these supraintestinal arteries give off two vessels, the omphalomesenteric arteries, which course back to the yolk, completing the embryonic circulatory pattern. The embryo at this stage is shown in Figure 8-22.

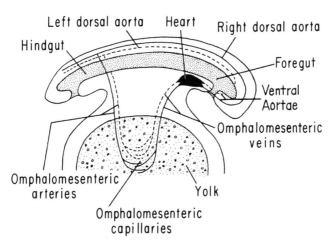

Figure 8-22. Early circulatory pattern in the teleost embryo.

Later Development of the Teleost

With the completion of the above development, the basic plan for most of the adult organ systems has been established. The subsequent development of specific organs within each of these systems is so very similar to that found in the frog, that unnecessary repetition would occur if it were included under both the teleost and the amphibian development.

The next chapter is complete to metamorphosis, and the student is referred to this chapter for a detailed account of the future development of the various organ systems.

"All scripture is given by inspiration of God, and is profitable for doctrine, for reproof, for correction, for instruction in righteousness: That the man of God may be perfect, thoroughly furnished unto all good works." II Timothy 3:16, 17.

Chapter **9**

The Embryology of the Frog

The class Amphibia presents an interesting embryological study. The amphibian egg has long been a target of study for many embryologists. This is due primarily to the easy accessibility of this egg. It is also due to the fact that it is easily maintained under average laboratory conditions. Because of the great interest in the amphibian egg, it is usually included in every course of vertebrate embryology. For that reason, the complete embryology of the amphibian egg will be given in this chapter. The type specimen chosen for this study is *Rana pipiens*, the common leopard frog.

Morphology of the Egg

The egg of the frog can be classified as mesolecithal and moderately telolecithal. The yolk is distributed toward the vegetal pole. In external appearance the egg presents a very dark animal hemisphere and an extremely light vegetal hemisphere (Fig. 9-1). The egg undergoes its first meiotic division within the oviduct of the female. When it is deposited in the water, it is in the middle of the second meiotic division. The egg is released directly into the water, and fertilization is external. The size of the egg is approximately 1.75 mm in diameter.

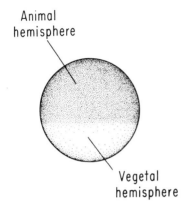

Figure 9-1. An egg of the grassfrog, *Rana pipiens*. This is a moderately telolecithal egg. The yolk is concentrated in the vegetal hemisphere.

Fertilization of the Egg

Shortly after the spring thaw the males and females are stimulated sexually by an increase in the amounts of gonadotropic hormones secreted by the pituitary gland. This results, in the female, in the process of ovulation. In the male the sexual stimulation is manifested by a sexual embrace termed amplexus. In this embrace the male frog mounts the female and encircles her trunk with his forelimbs. While in this position, the female sheds her eggs directly into the water, and the sperm from the male are deposited on top of them. The sperm usually enter the eggs immediately. In the frog fertilization is of the monospermy type.

The sperm enters the egg in the animal hemisphere. Once inside the egg, the sperm attempts to reach its destination, the female pronucleus. This, however, may require a change in its initial direction. The initial direction, i.e., the path taken by the sperm after entering the egg, is referred to as the penetration path. If the sperm must alter its direction in order to reach the female pronucleus, the new path is called the copulation path. In some cases the sperm need not alter its direction. In these cases the penetration and copulation paths would be identical. This can be seen in Figure 9-2. After the female pronucleus completes its second meiotic division, the fusion of the male and female pronuclei takes place. This completes the fertilization process.

When a sperm enters the egg, it disturbs the pigmented cytoplasm in the cortex of the egg and drags some of it along with it into the interior. This causes a surface shifting of the remaining pigmented material in the directions shown in Figure 9-3. The result is an area opposite the sperm entrance point which now has a slightly

The Embryology of the Frog 105

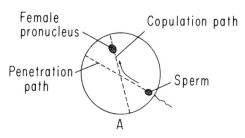

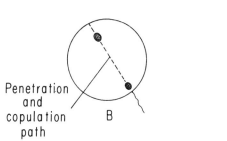

Figure 9-2. Possible sperm pathways in fertilization. (A) If the sperm must alter direction in order to fuse with the female pronucleus, the new route is called the copulation path. (B) In the event that the sperm need not alter its direction, the penetration and copulation paths are identical.

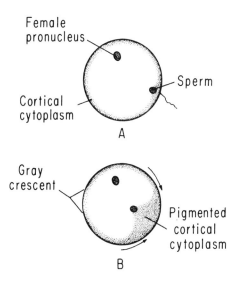

Figure 9-3. Formation of the gray crescent. When a sperm enters the egg (A), it disturbs the pigmented cortical cytoplasm, dragging some of it into the interior. This causes a surface shifting of the remaining pigmented cytoplasm (B). The result is an area opposite the sperm entrance point that now has a slightly decreased pigment content. This area is the gray crescent.

decreased pigment content. This resulting lighter pigmented area is called the gray crescent.

The animal pole, plus the sperm entrance point and the gray crescent, constitute the median plane of the embryo, and the first cleavage plane will bisect the gray crescent material.

Cleavage in the Amphibian Egg

As a result of the large amount of yolk in the vegetal hemisphere, cleavage in the amphibian egg is somewhat distorted. Although the entire egg cleaves, it does so unequally. Utilizing the usual terminology, this would be classified as cleavage of the holoblastic unequal type. The first cleavage plane runs from animal to vegetal pole through the gray crescent. This divides the egg into an equal left and right blastomere. The second cleavage plane is also oriented from animal to vegetal pole but at right angles to plane number one. The result is four equal blastomeres. The third cleavage plane is the one that is displaced. Instead of occurring near the meridian as it does in an isolecithal egg, it is displaced slightly toward the animal pole. It is at right angles to both planes one and two. This cleavage plane, therefore, results in four smaller animal blastomeres, the micromeres, and four larger vegetal blastomeres, the macromeres. The fourth cleavage plane is also oriented from animal to vegetal pole: it is, in reality, two cleavage planes, each one oriented at right angles to each other and at right angles to plane three. This divides each of the micromeres and macromeres into two cells. Therefore, the result of the fourth cleavage plane is eight animal micromeres and eight vegetal macromeres. The fifth cleavage plane, although irregular, is usually a double cleavage plane oriented parallel to plane three. This separates each of the animal micromeres and vegetal macromeres into two cells. The result is 16 animal micromeres and 16 vegetal macromeres. Cleavage, after this, is highly irregular and cannot be followed with any precision. Throughout the cleavage pattern, however, the animal cells remain small, while the heavy yolk-laden cells of the vegetal hemisphere are larger in comparison. These cleavage planes are summarized in Figure 9-4. The result of this unequal type of

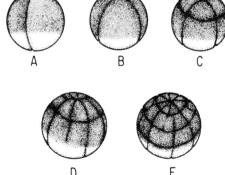

Figure 9-4. The first five amphibian cleavage planes. (A) The first cleavage plane. (B) The second cleavage plane. (C) The third cleavage plane. (D) The fourth cleavage plane. (E) The fifth cleavage plane. For a complete description of this cleavage pattern, see the text.

cleavage is a centrally located blastocoele which is displaced toward the animal pole. A sagittal section through an amphibian egg in the blastula stage is seen in Figure 9-5. It should be noted that the roof of the blastocoele is occupied by the smaller animal micromeres, whereas the floor of the blastocoele is occupied by the larger, yolk-laden vegetal cells.

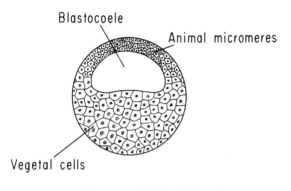

Figure 9-5. A sagittal section through a frog blastula. The roof of the blastula is composed of the smaller animal micromeres, whereas the floor is lined with the larger, yolk-laden, vegetal macromeres.

Fate Map of the Amphibian Blastula

Each part of the amphibian blastula has a prospective fate. It is possible by using vital stains (stains that mark portions of the embryo without having any deleterious effect) to follow the subsequent development of all parts of the amphibian blastula. As these vital stains mark the prospective fate of various parts of the blastula, fate maps can be accurately drawn. Figure 9-6 illustrates a fate map of an anuran blastula.

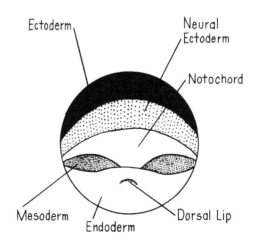

Figure 9-6. Fate map of a frog blastula.

If we use the dorsal lip of the blastopore as a point of reference, it can be seen that the area immediately surrounding the lip is presumptive endoderm. In the dorsal midline, working anteriorly, the surface of the blastula contains notochordal material, the presumptive neural area, and finally the ectoderm. Lateral to the notochord is the presumptive mesoderm.

Gastrulation

In the amphibian blastula, the blastoderm has been seen to contain the presumptive materials of the future endoderm, mesoderm, and ectoderm. During gastrulation, the endodermal and mesodermal portions move into the interior of the embryo. The process of gastrulation, therefore, consists of a shifting of groups of cells, creating new relationships and ultimately a triploblastic or three-germ-layered embryo. Figure 9-7 diagrammatically depicts the basic gastrulation movements.

The cells of the animal hemisphere still continue to grow more rapidly than those in the vegetal region, due to the relatively small amount of yolk present. The movement of these cells might be analogized to the flow of molasses after it has been poured onto the top of an orange. In that case, the molasses will flow down and around the surface of the orange. In the case of the blastula, the animal cells, reproducing at a fast rate, flow down and around the rest of the egg. This flowing movement of cells is known as epiboly. This epibolic movement continues around the entire surface of the egg, but it is interrupted at the area formerly occupied by the gray crescent. At this point, the downward-growing cells begin to turn in, i.e., involute. This point of involution is referred to as the dorsal lip of the blastopore. As involution continues, a new internal cavity is formed, replacing the original blastocoele. This cavity is the primitive gut or archenteron. It is also called the gastrocoele or cavity of the gastrula. The roof of the archenteron is composed of a thick layer of presumptive endoderm, notochord, and mesoderm. As involution continues, the archenteron grows in size, nearly obliterating the original blastocoele. The rest of the blastoderm continues to flow around the surface of the egg until it approaches the dorsal lip. Here, it also involutes, forming the ventral lip of the blastopore. The sides of the blastopore are also approached by the ectoderm. These are the lateral lips. If one looks into the area of the dorsal lip, the large yolk-laden cells can be seen through the darker ectoderm. This developmental state signifies the end of gastrulation, and it is called the yolk-plug stage. A sagittal section through the yolk-plug can be seen in Figure 9-7D.

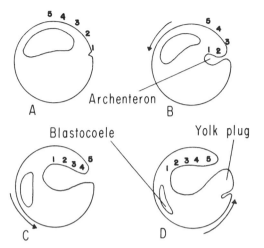

Figure 9-7. Gastrulation movements in a frog egg. As the ectoderm involutes around the dorsal lip, a new cavity, the archenteron (gastrocoele), is formed. The original blastocoele is "squeezed" out of existence. Once the ventral lip of the blastopore is formed, the yolk-laden cells seen through the blastoporal opening are called, collectively, the yolk plug.

The roof of the archenteron then delaminates, forming a definitive endodermally lined gut and a layer of cells located between the roof of the gut and the overlying ectoderm. This layer of cells is the chordamesoderm, so called because it is composed of both presumptive notochord and mesoderm. A cross section through the late gastrula is shown in Figure 9-8.

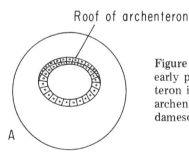

Figure 9-8. (A) Transverse section through an early postgastrula stage. The roof of the archenteron is mesendoderm. (B) Once the roof of the archenteron delaminates, a new layer, the chordamesoderm, is formed.

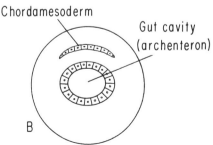

External Development to 48 Hours

The above knowledge of the relative positions of the various tissues of the early postgastrula embryo is extremely important for an understanding of the future development of the larva. The notochord lies in the median line of the anterior-posterior axis. Immediately above the notochord is the neural ectoderm, destined to become the central nervous system of the adult. This neural ectoderm is induced by the underlying notochord. Externally, the first indication of the development of the neural ectoderm is the appearance of two lateral neural ridges and a transverse neural ridge which connects the anterior ends of both lateral ridges. Between the two lateral neural ridges is a depression, the neural groove. This can be seen in Figure 9-9. As this neural development occurs, the embryo undergoes a period of elongation. The original gastrula was still spherical. However, within a short period of time, the embryo elongates to a size approximating 3 mm. In addition, the external indications of underlying organ development appear. Most of the external changes can be described as either elevations or depressions in the originallly smooth surface of the egg.

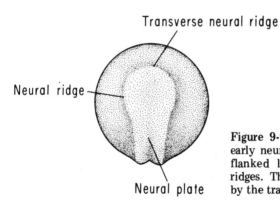

Figure 9-9. External appearance of an early neurula stage. The neural plate is flanked laterally by the two neural ridges. These are connected anteriorly by the transverse neural ridge.

The original neural ridges continue to fold up. These meet one another just anterior to the middle of the embryo. As this folding up continues in both an anterior and posterior direction, a prominent ridge is formed along the back of the embryo. Anteriorly, the area beneath the transverse folds begins to thicken appreciably, forming a shield. This shield is called the sense plate. It merges imperceptibly with the neural ridges posteriorly. Just posterior to the sense plate is another thickening, the gill plate. It is in this gill plate that the subsequent development of the gills will occur. An embryo, showing the neural tube, sense plate and gill plate, can be seen in Figure 9-10.

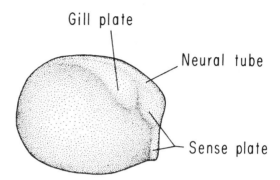

Figure 9-10. Lateral view of a late neurula.

The sense plate itself begins to undergo a period of internal differentiation. Externally, this appears as a pair of elevations dorsally, the forerunners of the eyes. These bulges are external indications of the developing optic vesicles. In the midline of the sense plate, oriented in a dorsal-ventral direction, is a depression, the stomadeal invagination. This separates the original sense plate into two halves. Each half is a mandibular arch. At the ventral extension of each mandibular arch, another invagination, the oral sucker or mucous gland, forms. Although these are transitional structures, they are very important because, after hatching, the larvae attach to rocks or plants by means of the sticky excretion that these mucous glands produce. Figure 9-11 shows the location and position of the optic

Figure 9-11. Anterior view of a late neurula.

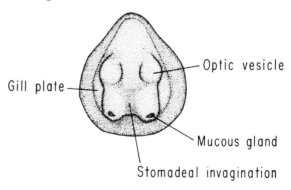

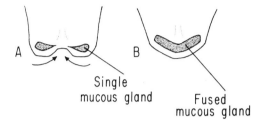

Figure 9-12. Mucous gland development. (A) A single mucous gland develops near the ventral end of each mandibular arch. (B) These soon grow toward one another until fusion occurs. This results in a single V-shaped mucous gland.

vesicles, stomadeal invagination, mandibular arches, and oral sucker or mucous gland. The original dual oral suckers soon begin to grow ventrally and eventually fuse with one another to form a single mucous gland. This mucous gland development can be seen in Figure 9-12.

The indentation between the sense plate and the gill plate is the hyomandibular cleft. Another change occurs in the posterior region of the gill plate as a depression, the forerunner of the fourth branchial cleft. Shortly after this, two new invaginations occur anterior to the fourth branchial cleft. These are the first and second branchial clefts. The third branchial cleft does not appear until much later in development. The solid areas between clefts are the branchial arches. Figure 9-13 demonstrates the position of these newly formed bran-

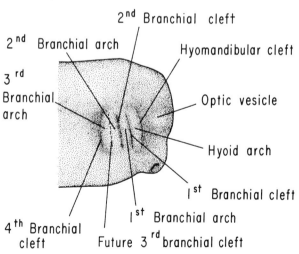

Figure 9-13. Head region, showing the position of the developing branchial clefts and arches.

chial clefts and arches. At this point, it is important to differentiate between the use of the terms "visceral arches" and "branchial arches." The term "visceral arch" is a strict anatomical term with no reference to function. The mandibular arch, therefore, is the first visceral arch; the hyoid, the second visceral arch; and the third, fourth, fifth, and sixth are numbered in sequence. If, however, the term "branchial arch" is used, the mandibular and hyoid arches are not included. A branchial arch is associated with respiration and usually bears a gill. The third visceral arch, therefore, becomes the first branchial arch.

Posteriorly, the blastopore, originally round and exposing the underlying yolk-laden endodermal cells, begins to narrow into a slit. As this narrowing process continues, a fusion of the lateral blastoporal lips takes place in the midregion of the slit. This creates two openings. The dorsal opening is the primitive streak; the ventral opening is the proctodeum (Fig. 9-14). The primitive streak of the anuran is homologous with the primitive streak of the avian embryo.

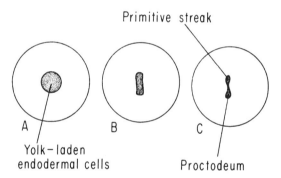

Figure 9-14. Development of the blastopore. (A) Originally the blastopore is round, exposing the underlying yolk-laden endodermal cells. (B) The blastopore narrows into a slit. (C) As the narrowing process continues, a fusion of the lateral blastoporal lips takes place in the midregion of the slit. This creates two openings, a dorsal primitive streak and a ventral proctodeum.

The form of the body after gastrulation is altered considerably, changing from an oval to a more typical larval shape. Posterior to the fourth branchial cleft, another thickening, the pronephros or head

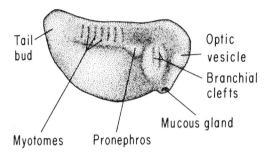

Figure 9-15. Lateral view of a 48-hour frog embryo, showing the most prominent structures visible externally.

kidney, makes its appearance. Throughout the body W-shaped myotomes begin to grow, dorsal to ventral (Fig. 9-15). All these changes take place within 48 hours after fertilization. Obviously, these external changes are only reflections of the tremendous amount of internal development that is taking place during the same period.

Internal Changes Within 48 Hours

Ectodermal Development

The ectoderm of the neural region is composed of an outer nonnervous and an inner nervous layer of ectoderm. This distinction is very important, because the derivatives of each of these layers are different. In the formation of the neural tube, the neural ridges begin to form (Fig. 9-16). As these elevate, they create a neural groove. These neural ridges soon approach one another in the dorsal midline and, ultimately, fuse. This creates a tube directed anterior to posterior—the neural tube. The cells of the nervous layer in the area of the fusion (the neural crest cells) begin to migrate away from their original dorsal position. These will form, in the future, five distinct structures in the embryo. These will be discussed later in more detail. At this point, however, it should be mentioned that they do form the spinal ganglia, cranial ganglia, autonomic ganglia, the cells of the adrenal medulla, and the chromatophores or pigment cells.

Anteriorly, the neural tube is expanded into a bulblike structure, the future brain. The last part of the neural tube to close completely, in the anterior end of the embryo, is the anterior neuropore. When this finally does close, it leaves a slight thickening. This

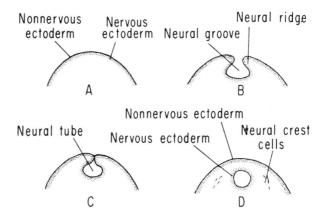

Figure 9-16. Development of the neural tube. (A) The ectoderm of the neural plate is composed of an outer nonnervous and an inner nervous layer of ectoderm. (B) The neural ridges begin to elevate, creating a neural groove. (C) These ridges approach and meet one another in the dorsal midline. (D) Ultimately the neural ridges fuse, forming a distinct neural tube. Note that the nervous ectoderm is on the outside of the tube. Cells of the nervous layer, in the fusion area, break loose and migrate. These are the neural crest cells.

thickening can be used as a landmark for helping to distinguish the various areas of the embryonic brain. As the embryo increase in length, the neural tube assumes a slightly concave orientation. In the head region a definite cranial flexure occurs. the most anterior part of the floor of the brain is called the tuberculum posterius. Opposite this tuberculum posterius, in the roof of the brain region and just dorsal to the closed anterior neuropore, is a thickened area. This dorsal thickening can also serve as a landmark for delimiting the brain areas.

Based on the above points of reference, three primary brain regions can be distinguished. If a line is drawn from a point just anterior to the tuberculum posterius to a point just anterior to the dorsal thickening, and another line is drawn from a point just posterior to the tuberculum posterius and just posterior to the dorsal thickening, the brain will be divided into three distinct regions. The most anterior is the prosencephalon or forebrain; the midregion is the mesencephalon or midbrain; and the most posterior region, that which is continuous with the spinal cord, is the rhombencephalon or hindbrain. These points of reference, together with the three primary brain regions, can be seen in Figure 9-17.

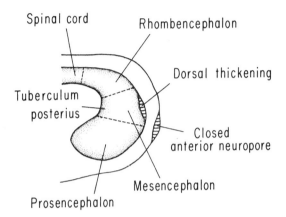

Figure 9-17. Sagittal section through the head region of a young embryo. The major regions of the brain are, at this stage, arbitrarily delimited by drawing two lines. The first is drawn from a point just anterior to the tuberculum posterius to a point just anterior to the dorsal thickening. The second is drawn from a point just posterior to the tuberculum posterius to a point just posterior to the dorsal thickening. In the figure these divisions are shown by dotted lines.

Shortly after the formation of the three brain regions, a ventral evagination in the prosencephalon occurs. This evagination is the infundibulum. As the infundibulum grows, an invagination from the roof of the stomadeum grows toward the infundibulum. This ectodermal ingrowth is Rathke's pocket. Rathke's pocket soon pinches off from the roof of the stomadeum, forming a vesicle which continues to move toward the infundibulum. Eventually, this fuses to the infundibulum to form the rudimentary pituitary gland. This pituitary development is depicted in Figure 9-18. Rathke's pocket forms the pars distalis, pars intermedia, and pars tuberalis of the adult gland, whereas the infundibulum forms the pars nervosa.

In the lateral region of the prosencephalon, a pair of evaginations begin to develop. These are the forerunners of the sensory organs, the eyes. These optic primordia continue to grow laterally until they press upon the overlying skin ectoderm. Once in this position, they begin to bulge in the area of contact. This bulging of the

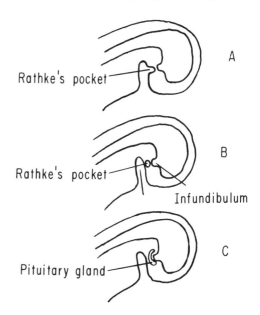

Figure 9-18. Pituitary development. (A) Shortly after the formation of the three brain regions, a ventral evagination of the prosencephalon occurs. This is the infundibulum. An invagination from the roof of the stomadeum also develops. This is Rathke's pocket. (B) Rathke's pocket soon pinches off to form a vesicle which migrates toward the infundibulum. (C) Rathke's pocket ultimately fuses with the infundibulum to form the pituitary gland (hypophysis).

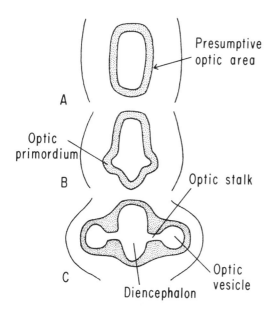

Figure 9-19. Development of the optic vesicles. (A) Cross section through the diencephalon prior to the appearance of the optic primordia. The arrow points to the presumptive optic area. (B) Same section shortly after the appearance of the optic primordia. (C) The optic primordia bulge at their distal ends creating optic vesicles. The thin portion attached to the diencephalon is the optic stalk.

distal portion of the original optic primordia divides it into a connecting optic stalk and an enlarged optic vesicle (Fig. 9-19). This optic vesicle causes the bulges that are apparent in the head region at this stage.

In the skin, in the area of the hindbrain, a pair of thickenings, which at this stage are called the auditory placodes, begin to develop. These auditory placodes will eventually develop into the auditory or ear vesicles of the adult frog.

In a similar fashion, the head ectoderm just beneath the prosencephalon begins to thicken as nasal placodes. These nasal placodes will eventually deepen to form the nasal chambers, or choanae, of the adult frog.

These are the major ectodermal changes that occur during the early 48-hour period of frog development. The development of each of these structures will be discussed when we survey the later larval development.

Endodermal Development

The primary gut cavity, as a result of the anterior-posterior extension of the embryo, can now be delimited into three primary regions (Fig. 9-20). That portion of the gut which extends anterior to

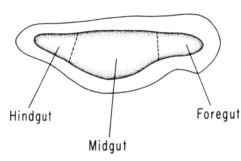

Figure 9-20. Regions of the gut. The dotted lines represent the anterior and posterior limits of the original yolk-laden cells.

the yolk mass is the foregut. That portion which extends posterior to the yolk mass is the hindgut. The remaining portion and, at this stage the largest portion, that which contains the yolk, is the midgut. Each of these gut areas undergoes an extensive period of development and ultimately gives rise to a number of adult structures. The foregut develops into the mouth, pharynx, esophagus, stomach, liver, gallbladder, and pancreas. The midgut develops into the small and large intestine, and the hindgut develops into the rectum and cloaca.

The most extensive development in the early embryo occurs in the region of the foregut. A ventrally directed oral evagination begins to push toward an ectodermal invagination, the stomadeum. As these two developmental pouches meet one another, a double-layered tissue, the oral plate, forms. After hatching, this ruptures to form the mouth cavity. It can be seen, therefore, that the definitive mouth is composed of ectoderm externally and endoderm internally. In the region of the gill plate, a series of evaginations occur in the lateral walls of the pharynx. Arising at approximately the same time are the visceral grooves discussed earlier in the description of the external anatomy of the early larva. These depressions create a series of isolated arches from the originally solid body wall. The first arch is the mandibular arch, the second is the hyoid. The first evagination from the endoderm and the invagination from the ectoderm form the hyomandibular groove. The development of the hyomandibular groove as well as that of all the more posterior arches is shown in Figure 9-21.

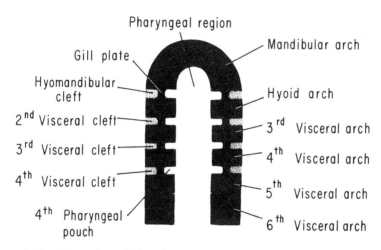

Figure 9-21. Formation of the visceral arches and grooves. The invaginations from the ectoderm meet the evaginations of the endoderm to form the gill plates. When these rupture, a gill slit is formed. The mandibular arch is the first arch, and the hyoid is the second arch.

The only other primordium to develop at this time in the foregut region is a ventrally directed outgrowth immediately anterior to the midgut region. This ventrally directed outgrowth, although called a liver primordium, is actually the anlage for the liver, bile duct, and gallbladder.

The midgut does not show any extensive development during this early period. The yolk, however, continues to be utilized, and the diameter of the midgut region decreases directly with the yolk usage. The hindgut shows a ventrally directed evagination similar to the evagination in the oral region. This ventrally directed outgrowth meets an invagination from the ectoderm, the proctodeum. Eventually these meet, forming the cloacal plate. This eventually ruptures to form the cloacal opening.

Mesodermal Development

The mesoderm in the head region is in the form of loosely aggregated cells called, collectively, mesenchyme. These mesenchymal cells, in the area of the developing gill pouches, form the internal structures in gill arches. These were described under endodermal development.

The only other extensive mesodermal development to occur during the first 48 hours is the growth of the lateral mesoderm. The lateral mesoderm is composed basically of a dorsal, medial enlargement, the somite or epimere; a middle section, the nephrotome or mesomere; and a lateral split area, the hypomere. This hypomere is sometimes referred to as the lateral plate mesoderm. The epimere, shortly after its development from the roof of the archenteron, begins to segment into distinct entities called somites. These appear first anteriorly and develop progressively posteriorly. These somites develop within themselves a cavity, the myocoele. This divides the somite into an outer and an inner layer. The outer layer is the dermatome, sometimes called the cutous plate. This develops into the dermis of the skin. The inner layer is the myotome. This is so named because it will ultimately develop into many of the striated muscles of the back. On the most medial portion of the myotome, cells bud off to form the cartilaginous sheath around the notochord and neural tube. This area of budding cells is referred to as the sclerotome (Fig. 9-22).

Lateral to the somite is the very thin nephrotome or mesomere. Shortly after its formation, it also develops a cavity within it, the nephrocoele. This nephrocoele and nephrotome are the forerunners of the pronephric kidney of the larval amphibian. This kidney development will be followed in the later stages.

Lateral to the mesomere is the lateral plate mesoderm or hypomere. This is split into an inner splanchnic and an outer somatic layer. The somatic mesoderm and ectoderm are referred to as the somatopleure. The splanchnic mesoderm and the closely applied endoderm of the gut are collectively called the splanchnopleure. The cavity between the two is the coelomic cavity. A cross section through an embryo of this stage, showing the development of the lateral mesoderm, can be seen in Figure 9-23.

Figure 9-22. Partial cross section through the somite area. The cavity within the somite is the myocoele. This separates the somite into an outer dermatome and an inner myotome. The sclerotomal cells which bud off from the medioventral portion of the somite will form the basic vertebral components around the neural tube and notochord.

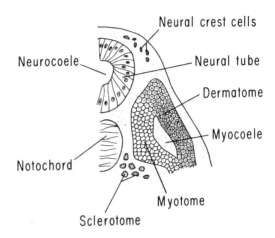

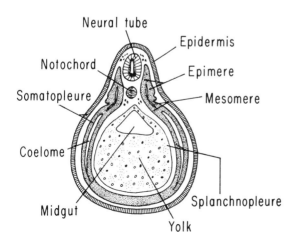

Figure 9-23. Typical cross section through the midgut region of a 48-hour frog larva. The major structures are shown.

Later Development

The entire period of later amphibian development can be considered as occurring in three relatively distinct stages. The first is that period prior to hatching. During this time, the embryo is still contained within its jelly envelopes. It is not a free-living form. The second period, the posthatching period, is that period of development which occurs after the embryo breaks from its shell. During the first part of this period, the embryo is still not completely free-living. It attaches itself to the sides of rocks and leaves or, in an aquarium, to the sides of the glass. Shortly after this, the embryo does become a free-living form. It is during this stage that the term "tadpole" is correctly applied to the developing embryo. This tadpole stage lasts from a few months to two years, depending on the species of frog. The third stage is actually a period of transformation from the tadpole to the adult frog. During this time, the various organ systems undergo a rather drastic transformation. This transformation alters the morphology and physiology of the aquatic tadpole to adapt it to the terrestrial existence of the adult frog. This period of transformation in amphibians is referred to as metamorphosis.

External Development

During the prehatching stage, the embryo continues to elongate. The primary area of elongation is the tail itself. This elongation of the tail makes it possible to divide the body arbitrarily into an anterior head, a middle belly, and a posterior tail region. The extension of the tail, posteriorly, includes certain embryonic structures which already occur within the body proper. These structures—the notochord, arteries, veins, myotomes, epidermis, and dermis—can now be seen in a cross section of this elongated tail (Fig. 9-24). The belly region is distended ventrally, for it is in this area that a large quantity of yolk still persists. It must be remembered that all the nutrition for the metabolically active cells of the embryo must come from this yolk source, since no external feeding is done prior to hatching.

The myotomes, which were present in the early embryo as V-shaped depressions in the lateral body area, now continue to develop posteriorly. As the tail elongates, these myotomes are conspicuous in the caudal area. They appear as a series of V's, with the apex of the V directed anteriorly.

The head region undergoes the most drastic development during the early stages. The most obvious protuberance in this area is the

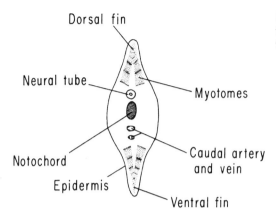

Figure 9-24. Cross section of the amphibian tail, showing the major structures.

optic vesicle. As the optic vesicles continue to expand laterally from the diencephalon, they impinge upon the underlying ectoderm, causing the extensive protuberance on each side of the head. In the midventral line, the stomadeal invagination continues to deepen. At the dorsal end of this invagination, the stomadeum proper develops. This continues to invaginate toward the endoderm of the gut and, in a short time, will rupture to form the mouth or oral cavity. The sense plate on each side of the stomadeal invagination also undergoes a period of external differentiation. A pit develops on each sense plate. This pit is the olfactory pit, the forerunner of the external nares.

Laterally, the gill plate continues to differentiate. A new cleft, the third branchial cleft, now makes its appearance. It is located between the original second and fourth branchial clefts. Since the embryo has now reached a stage where the diffusion of oxygen through the skin is no longer sufficient to satisfy the metabolic needs of the body cells, new structures begin to develop to take over the function of respiration. These appear as a series of fine, fingerlike filaments on the branchial or gill arches (Fig. 9-25). The external gills make their appearance, in these early prehatching stages, on the first and second branchial arches. They are highly vascularized and allow for an adequate exchange of oxygen and carbon dioxide be-

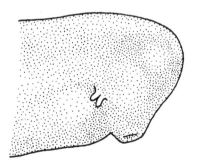

Figure 9-25. Head of a young larva showing the appearance of external gills.

tween the embryonic blood and the external water environment. Posterior to the gill plate, the pronephros, or head kidney, continues to enlarge and becomes a rather conspicuous structure. At this stage the embryo begins to undergo a series of body movements, which indicate a functioning muscular system. These body movements, together with a rupture of the egg shell, allow the embryo to escape or hatch. This hatched embryo is now properly referred to as a tadpole. Although the size of the tadpole at hatching will vary from species to species, it is approximately 8 mm in length in *Rana pipiens*. A newly hatched tadpole is shown in Figure 9-26.

Figure 9-26. External appearance of a larva shortly after having broken out of the egg capsule.

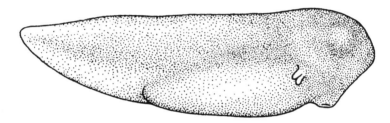

Posthatching External Development

When the embryo breaks from its confining shell, it swims to a nearby rock or, in an aquarium, to one of the glass sides. Here, it attaches by means of its mucous glands. It is still incapable of feeding because of the lack of a complete digestive tract. The stomadeum and proctodeum are not open at this stage. The yolk, still present in the midgut region, is the sole source of nutrition. Soon, however, these openings do form and the embryo becomes a free-swimming tadpole. Its diet, during the tadpole stage, is primarily vegetative, although frequently the tadpole will be found devouring decaying animal meat. The body shape, in general, becomes altered from the typical embryonic condition. The body, itself, becomes compressed in a dorsal-ventral direction, whereas the tail is compressed laterally. The tail also continues to grow at a rather rapid rate, until it becomes approximately twice the length of the body. The embryo, at the time of hatching, is brown in color. This changes, gradually, to a bichromatic condition, where the dorsal side of the embryo is green and the ventral surface a white or pale yellow.

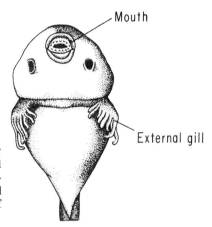

Figure 9-27. Ventral view of an early posthatching larva. Note that the original stomadeal invagination has developed into a round mouth, which is surrounded by circular jaws containing many rows of sharp papillae.

Anteriorly, the mucous gland atrophies shortly after the tadpole becomes free-swimming. The stomadeal invagination develops into a round mouth (Fig. 9-27), which is surrounded by circular jaws and lips which are covered by many rows of sharp papillae. The olfactory pit continues to deepen. In the area of the optic protuberance, an invagination of the overlying ectoderm over the central portion of this protuberance signifies the appearance of the embryonic lens. Posterior to the optic protuberance is another invagination, the otic pit. This will become the auditory sense organ of the adult.

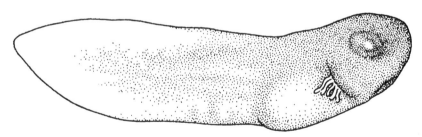

Figure 9-28. External view of an early free-swimming tadpole.

The gill placode undergoes still further development. The third and fourth visceral arches develop external gills. A rudimentary gill also develops on the fifth arch. This continuation of external gill development is consistent with the tadpole's increasing oxygen needs. Water circulates in through the mouth and out through the branchial clefts; over the external gills. An early free-swimming tadpole can be seen in Figure 9-28.

A new structure, the operculum, begins to develop from the posterior border of the hyoid arch. The operculum continues to grow posteriorly, covering the external gills. On the right side of the body, the operculum fuses with the body ectoderm. On the left side of the body, however, it remains open for some period of time. This opening is the spiracle (Fig. 9-30). A space is thereby created between the original body wall and the opercular covering. This is the gill chamber (Fig. 9-29).

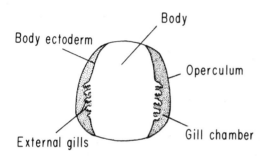

Figure 9-29. Diagrammatic representation of the gill region after the development of the opercular folds. The gill chamber communicates with the external environment through the spiracle (See Fig. 9-30).

The ventral body wall is very thin, and through it the long, coiled intestine is evident. At the point where the body and tail join, the proctodeal invagination can be seen. This is the primordium of the cloacal opening. When both the stomadeum and proctodeum are fully formed, the digestive tube is complete. Feeding on external vegetative and decayed animal matter can then occur.

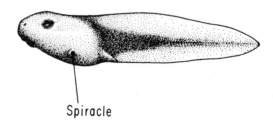

Figure 9-30. Full-fledged tadpole. The spiracle remains open only on the left side.

The tail develops a continuous fin around both its dorsal and ventral surfaces. After all these changes have taken place, the embryo is a full-fledged tadpole (Fig. 9-30). It remains in this condition for two to three months in *Rana pipiens.* In some of the larger frogs, such as *Rana catesbiana,* the bullfrog, the embryo goes through an entire winter and part of the following summer before additional changes occur.

External Changes at Metamorphosis

At the end of the first summer in *Rana pipiens* larva, and at the end of the second summer in some of the larger frogs, such as *Rana catesbiana* (the bullfrog), a transformation takes place which transforms the swimming larva, or tadpole, into a miniature frog. This transformation is called metamorphosis. This change is truly remarkable when one considers that it transforms an aquatic animal into one whose organ systems are essentially adapted to a terrestrial existence. This metamorphosis requires either an alteration in, or a replacement of, most of the larval organ systems. The complete metamorphic cycle is under the control of throxin, a hormone produced by the thyroid gland. If iodine, one of the major components of thyroxin, is withheld from the diet of the tadpole, metamorphosis will not occur. In the laboratory, metamorphosis can be accelerated by feeding the tadpole either thyroxin or iodine.

One of the more drastic changes occurs in the respiratory system. Throughout the larval life respiration is that of the typical aquatic form. The oxygen which is dissolved in the water passes through the thin linings of the external gills. Once the animal becomes terrestrial, these gills can no longer function, and respiration is accomplished by means of a new respiratory system, with the lungs serving as the primary respiratory organ. During the metamorphic changes, therefore, the gills can be seen to close, and the opercular slit, which served as the point of exit for circulating gill water, also is closed.

The horny jaws which were found within the circular mouth of the tadpole were used primarily for scraping vegetation from rocks or from the sides of the aquaria. During metamorphosis, these jaws widen considerably, altering the shape of the mouth from a circular or oval opening to a large slit. The horny teeth which were characteristic of the tadpole are replaced by true maxillary teeth. The circular gut which appeared as a coiled structure, and which could be seen through the ventral abdominal wall, now shortens considerably. This is consistent with a change in the nutritional habits of the animal as it passes through the stages of metamorphosis. During its larval existence it was essentially herbivorous; in the adult form the frog is basically carnivorous. These diametrically opposed types of nutrition require changes throughout the digestive tract. The change in the length of the gut is one that can be seen externally.

Shortly after the onset of metamorphosis, a pair of buds appear at the junction of the tail with the body. These are the posterior limb buds. These continue to grow and form complete hind limbs approximately halfway through the metamorphic process. The anterior limbs also develop, but their development is withheld from view because they are covered by the opercular membrane. Shortly before the metamorphic process is completed, the anterior limbs push through the opercular membrane, thereby creating a true four-footed form.

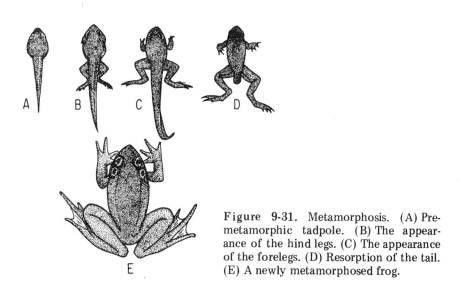

Figure 9-31. Metamorphosis. (A) Premetamorphic tadpole. (B) The appearance of the hind legs. (C) The appearance of the forelegs. (D) Resorption of the tail. (E) A newly metamorphosed frog.

Still another major change is the loss of the tail. The adult frog is tailless, while the larval tadpole depends greatly on its tail for locomotion. This shortening of the tail is gradual and occurs throughout the metamorphic process. In fact, the newly metamorphosed frog still has remnants of the original tail. All these external changes reflect rather extensive internal developmental changes. These will be covered in the pages to follow. The basic external changes that are evident throughout the metamorphic process are illustrated in Figure 9-31.

Posthatching Internal Development

Ectoderm

Development of the Brain

The early amphibian brain is composed of three primary brain portions. Most anteriorly is the prosencephalon; posterior to this is the mesencephalon; and most posteriorly is the rhombencephalon. As development proceeds, the prosencephalon further subdivides into two distinct brain regions, the anteriorly located telencephalon followed immediately by the diencephalon. The brain of the adult frog is, therefore, a four-part brain composed, in an anterior-posterior sequence, of the telencephalon, diencephalon, mesencephalon, and rhombencephalon.

The most anterior brain portion, the telencephalon, contains within it a cavity, the telocoele. The telencephalon, during the course of development, splits into a left and right portion. Each part is then called a cerebral hemisphere. Each cerebral hemisphere buds off anteriorly to form an olfactory lobe. This is fairly prominent in the frog. Although the origin of the olfactory lobes is dual, that is, each arising from a cerebral hemisphere, they soon fuse to form a single, anteriorly located olfactory lobe. The olfactory nerves, the first cranial nerves, connect the olfactory lobes with the olfactory epithelium, which is derived from the nasal placodes. The telocoele also subdivides as the cerebral hemispheres are formed. These cavities are now referred to as brain ventricles. That portion of the telocoele found within the left cerebral hemisphere is the first brain ventricle; that found within the right cerebral hemisphere is the second brain ventricle. These are connected to one another and to the third brain ventricle by the foramen of Munro.

The diencephalon is located immediately posterior to the telencephalon. Its cavity, the diocoele, is the third brain ventricle. Laterally, as has already been seen, the optic vesicles form as evaginations from the diencephalic wall. These continue to grow, forming the eyes of the adult. Dorsally, an evagination, the epiphysis, forms. This forms a gland which is homologous to the pineal gland of the higher vertebrates. The floor of the diencephalon is not smooth but contains a depression, the optic recess, which is followed by a thickened area, the optic chiasma. It is within this optic chiasma that the optic nerves, the second cranial nerves, are found. These nerves, coming from the sensory portion of the eye, cross on their way to the brain in the optic chiasma. Another ventrally located structure in the

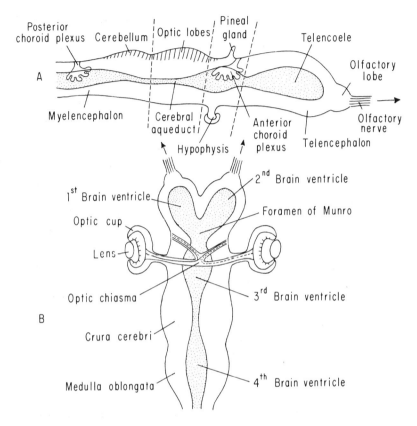

Figure 9-32. Embryonic structure of the frog brain. (A) A midsagittal section. (B) A frontal section.

diencephalon is the infundibulum. This infundibulum grows toward the developing Rathke's pocket from the stomadeum and eventually fuses with it to form the pituitary gland. Throughout the period of development and throughout adult life, this pituitary gland retains its connection to the ventral aspect of the diencephalon. Dorsally, an ingrowth from the roof of the diencephalon, the anterior choroid plexus, forms. This is a vascular network which is so extensive that it pushes into the cavities of the first, second, and third brain ventricles.

The mesencephalon develops extremely thick walls. These lateral thickenings are the crura cerebri and contain primarily nerve fibers which serve to connect various levels of the brain stem. The dorsal thickening continues to grow and eventually forms a pair of prominent structures, the optic lobes. These optic lobes contain reflex centers of vision. Since they form a two-lobed body, the term "corpora bigemina" is frequently applied to them. The cavity of the mesencephalon is narrowed because of the thickened development of the wall and roof. This mesocoele becomes the cerebral aqueduct (the aqueduct of Sylvius). This aqueduct connects the third and fourth brain ventricles.

The rhombencephalon is, in higher forms, subdivided into two distinct brain regions. In the frog, however, only the suggestion of a subdivision occurs. Anteriorly, in the rhombencephalon, a dorsal thickening begins to develop. This dorsal thickening is the cerebellum. Posteriorly, the walls and floor become the medulla oblongata. The cavity of the rhombencephalon, the rhombencoele, becomes the fourth brain ventricle which serves to connect the central canal of the spinal cord with the cerebral aqueduct of the mesencephalon. Growing from the roof of the rhombencephalon is a structure similar to the vascular ingrowth of the diencephalon. This vascular structure, the posterior choroid plexus, is rather extensive. In Figure 9-32 the developing frog brain can be seen. Most of the structures mentioned are located in their approximate position.

The Spinal Cord

That portion of the original neural tube which does not develop into brain becomes the spinal cord of the adult. As a result of its method of formation, the neurocoele, or cavity of the neural tube, is lined with cells that were originally epidermal. These cells are non-nervous and do not contribute to the formation of any of the nervous elements of the frog. They are referred to as ependymal cells. The outside of the spinal cord is derived from the original nervous ectoderm, that which was lying beneath the original epidermis. This nervous ectoderm is destined to form both neuroblasts (embryonic nerve cells) and neuroglial cells (embryonic supporting cells of the nervous system). Both the neuroblasts and neuroglia form the gray matter of the cord. As these neuroblasts begin to develop, by sending out processes (axons) which ascend or descend in the spinal cord, a new layer is seen. This is the white matter. This white matter is composed basically of cross sections of the ascending and descending axons. Figure 9-33 illustrates the development of the adult spinal cord from the embryonic neural tube.

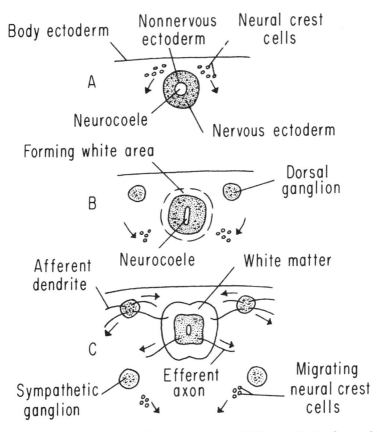

Figure 9-33. Development of the spinal cord. (A) The method of neural tube formation results in a neurocoele lined with the original outer body ectoderm. This nonnervous ectoderm is surrounded by nervous ectoderm containing both presumptive neurons and neuroglial cells. The neural crest cells begin migrating away from their site of origin. (B) Some of the neural crest cells aggregate dorsolateral to the spinal cord (Neural tube) as the dorsal ganglia. These cells are the progenitors of the afferent (sensory) neurons. The neurons within the spinal cord send processes to higher and lower levels of the chord. These nerve processes form the white matter. Some of the original neural crest cells continue to migrate ventrally. (C) The sensory neurones within the dorsal ganglion send axonic processes towards the spinal cord and dendritic processes to the periphery. The neurons in the ventral portion of the original neural tube send axonic processes toward the periphery. The neural crest cells aggregate ventrolaterally as the sympathetic ganglia. Others continue to migrate to other sites in the body where they will form chromatophores, adrenal medullary cells, or parasympathetic ganglia.

The Peripheral Nervous System

Spinal Nerves

The neural crest cells, which were derived from the closure of the neural folds, lie initially in a continuous band on either side of the neural tube. Soon, however, they become metamerically arranged as the result of the enlargement of the developing somites. Some of these neural crest cells aggregate dorsolaterally near the spinal cord to form the dorsal ganglia (Fig. 9-33). These neuroblasts differentiate into the sensory cells of the adult frog. Each of the sensory cells develops two processes: an axon and a dendrite. The axon begins to grow from the dorsal ganglion toward and into the dorsal-lateral wall of the spinal cord. The dendrites grow toward the skin to become the sensory or afferent components of the spinal nerve.

Neuroblasts in the ventral portion of the gray matter also develop axons and dendrites. The dendrites remain within the gray matter, while the axons emerge from the spinal cord and grow peripherally toward the skin and muscles. These axonic fibers become the efferent or motor components of the spinal nerve and form the ventral root.

Paired spinal nerves are formed throughout the length of the original neural tube. However, during metamorphosis the tail is resorbed. Those spinal nerves originally formed in the vicinity of the tail subsequently degenerate. Consequently, only ten remain after this metamorphic loss of the tail. The first spinal nerve is called the hypoglossal nerve, the second is the brachial nerve. The first, second, and third contribute to a plexus, the brachial plexus, which innervates the forelimb. The seventh, eighth, and ninth nerves, which innervate the leg, form the lumbosacral or sciatic plexus. The tenth nerve and fibers from the ninth form the ischiococcygeal plexus. These nerves innervate the urogenital organs. The distribution of the nerve fibers for each of these ten spinal nerves is as follows on page 134.

Between the endings of the sensory axons in the dorsal region of the gray matter and the dendrites of the motor neurons in the ventral portion there exists an expanse that must be bridged if these nerves are to be connected functionally to one another. This bridge is accomplished by means of association neurons, which are derived from neuroblasts in the original dorsal region of the neural tube. These neuroblasts send short dendritic processes toward the axon of the sensory neuron and a long axon which courses through the gray

Spinal Nerve Origin and Distribution

Number	Place of Emergence	Distribution
1	Posterior to the first vertebrae	Anteriorly to the tongue. Posteriorly, it merges with the second and third spinal nerves to form the brachial plexus
2	Posterior to the second vertebrae	Together with the first and third nerves it forms the brachial plexus. The brachial nerve, which innervates the forelimb and shoulder, arises from this plexus
3	Posterior to the third vertebrae	Brachial plexus and posterior shoulder region
4	Posterior to the fourth vertebrae	Skin and muscles of the anterior trunk region
5	Posterior to the fifth vertebrae	Skin and muscles of the middle trunk region
6	Posterior to the sixth vertebrae	Skin and muscles of the posterior trunk region
7	Posterior to the seventh vertebrae	Lumbosacral (sciatic) plexus. A few fibers are distributed to the posterior trunk muscles
8	Posterior to the eighth vertebrae	Lumbosacral (sciatic) plexus. This plexus is formed by the seventh, eighth, and ninth spinal nerves. The sciatic nerve, which innervates the posterior limb, arises from the plexus
9	Posterior to the ninth vertebrae	Lumbosacral (sciatic) plexus. A few fibers join with the tenth nerve to innervate the viscera in the posterior portion of the peritoneal cavity
10	Foramen in the urostyle	Visceral organs in the posterior region of the peritoneal cavity

matter toward the ventral area. Here, they connect or synapse with the dendrites of the motor neurons, completing a functional reflex arc (Fig. 9-34).

The Cranial Nerves

The counterparts to the spinal nerves in the brain region are called cranial nerves. There are ten pairs of cranial nerves in the frog. These are derived from three primary sources: (1) from neural crest cells, (2) from neuroblasts located within the developing brain, and (3) from parts of the nervous ectoderm that are associated with the

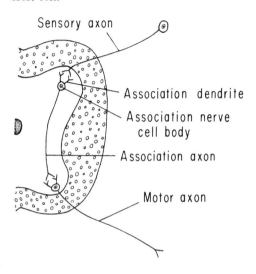

Figure 9-34. Diagram of a three-neuron reflex arc. The association and motor neurons are derived from the original neural tube. The sensory neuron is a differentiated neural crest cell.

Cranial Nerve Origin and Distribution

Number	Name	Origin	Distribution	Type
1	Olfactory	Olfactory lobe	Lining of nose	Sensory
2	Optic	Posterior floor of cerebral hemispheres	Retina of eye	Sensory
3	Oculomotor	Midbrain	Inferior oblique Superior rectus Inferior rectus Internal rectus	Motor
4	Trochlear	Midbrain	Superior oblique muscle of eye	Motor
5	Trigeminal	Medulla	Muscles of face, tongue, and jaw	Mixed
6	Abducens	Medulla	External rectus	Motor
7	Facial	Medulla	Muscles of face	Mixed
8	Auditory	Medulla	Ear	Sensory
9	Glossopharyngeal	Medulla	Pharynx and tongue	Mixed
10	Vagus	Medulla	Visceral organs	Mixed

differentiating sense organs of the head. If a cranial nerve is composed only of afferent fibers, it is a sensory nerve. The first, second, and eighth cranial nerves are of this type. Others possess only efferent fibers and are called motor cranial nerves. The third, fourth, and sixth nerves are of this type. Finally, there are those that, like the spinal nerves, contain both sensory and motor fibers. These are the mixed nerves. The fifth, seventh, ninth, and tenth cranial nerves are mixed. The ten cranial nerves and the origin and distribution of each within the head of the frog are as shown on page 135.

The Autonomic Nervous System

Another portion of the nervous system which has not been described is the autonomic nervous system. This is the motor portion of nerves going to the visceral organs. The autonomic nervous system is subdivided into two distinct systems. Those autonomic nerves that arise in the thoracic and lumbar regions of the spinal cord comprise the thoracolumbar or sympathetic system. Those nerves that originate in either the brain or sacral region of the spinal cord are collectively called the craniosacral or parasympathetic system. The basic difference between the autonomic nerves and a typical spinal nerve is that the motor neuron in the spinal nerve goes directly from the spinal cord to the muscle or structure being innervated. In the case of the autonomic nervous system, the motor nerve from the spinal cord or brain goes first to an autonomic ganglion, where it synapses with another motor nerve which carries the impulse to the visceral organ. The origin of both the sympathetic and parasympathetic ganglia is from neural crest cells. The sympathetic ganglia are found ventrolateral to the spinal cord. They are arranged segmentally and form a sympathetic trunk. The parasympathetic ganglia are located some distance from the spinal cord, usually on, in, or near the structure being innervated. Figure 9-35 shows the relationship of these autonomic fibers to the spinal cord.

The Eye

The eye originates as a lateral evagination of the diencephalon. It begins first as a vesicle, which later invaginates to form an optic cup. This optic cup lies in close proximity to the head ectoderm and induces in this ectoderm a lens (Fig. 9-36). The connection of the optic cup to the brain is the optic stalk. The simple act of invagination creates a number of definitive structures in the adult eye. The

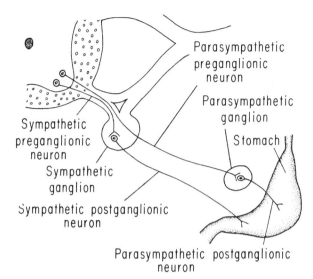

Figure 9-35. Diagram showing the relationship of the sympathetic and parasympathetic fibers to the central nervous system. For convenience, both sets of fibers are shown leaving this section of the spinal cord. In reality, the parasympathetic preganglionic fibers originate only in the brain and sacral region of the cord, whereas the sympathetic fibers originate in the thoracic and lumbar regions.

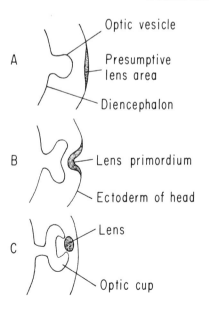

Figure 9-36. Development of the amphibian eye. (A) The eye originates as a lateral evagination of the diencephalon. (B) This optic vesicle then invaginates to form a double-walled optic cup. The optic cup induces the overlying ectoderm to invaginate as the lens primordia. (C) The lens vesicle breaks from the ectoderm and locates itself within the optic cup.

invaginated layer will become the retina; the original outer layer, the pigmented coat. The lateral opening is the pupil. Ventrally, the cup is not complete. The indentation which remains is the choroid fissure. The development of these eye parts can be seen in Figure 9-37.

Once the lens has formed, the overlying ectoderm closes over and differentiates to become the cornea. The space between the lens and the retina fills with a fluid secreted by the retinal cells. This fluid is the vitreous humour. The cells of the retina further differentiate into the rods and cone of the adult. Nerve processes from these cells grow to the brain through the optic stalk. This composite of nerve fibers is the second cranial nerve (the optic nerve) which has already been discussed.

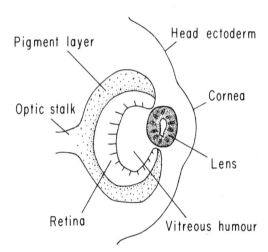

Figure 9-37. Continuation of the eye development seen in Figure 9-36. The connection of the optic cup to the brain is the optic stalk. The original outer layer of the optic cup is the pigmented layer, whereas the invaginated portion becomes the sensory retina. The ectoderm overlying the eye differentiates into the cornea.

The Ear

Ear development begins as a thickening of the lateral head ectoderm in the area opposite the rhombencephalon region of the brain. This ectodermal thickening, the auditory placode, invaginates to form a vesicle, the auditory sac (Fig. 9-38). Shortly after its formation, a membrane forms which separates the original auditory sac into a medial utricle and a lateral saccule. The three semicircular canals develop from the utricle. The saccule develops into the hearing portion of the ear. It develops a connection to the brain through the eighth cranial nerve, the auditory nerve. As this internal differentiation of the auditory sac is taking place, the covering ectoderm of the head becomes specialized as the tympanic membrane. There is no external ear in the frog.

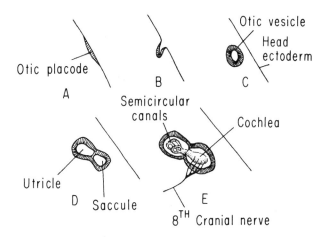

Figure 9-38. Development of the amphibian ear. (A) The first appearance of an ear is a thickened area of ectoderm opposite the rhombencephalon. (B) This auditory or otic placode invaginates to form the otic vesicle seen in C. (C) This original vesicle becomes separated into a medial utricle and a lateral saccule. (E) The utricle develops within it the semicircular canal system. The saccule differentiates into the hearing portion of the ear. These developing structures are connected to the central nervous system via the eighth cranial (auditory) nerve.

The Nasal Chambers

The olfactory placodes have already been observed as thickenings in the sense plate. These invaginate to form the olfactory pits. The external openings are the external nares. These continue to grow inward until they break through into the oral chamber as the internal nares. The resulting chambers, which connect the external and internal nares, are the choanae. They are lined with an epithelium which contains some neuroblasts. These neuroblasts differentiate into nerve cells which send axons into the olfactory lobe of the brain. These fibers compose the first or olfactory nerve.

Other Ectodermal Derivatives

The ectoderm forms other structures in addition to the nervous elements mentioned. The entire skin covering the frog's body is of ectodermal origin. Any invaginations occurring in the skin, such as the stomadeum and proctodeum, therefore, are also of ectodermal origin. The neural crest cells have already been observed contributing to the nervous ganglia. Other neural crest cells migrate to the site of development of the adrenal gland. Here they form the center, or medulla, of the adrenal gland. Still other neural crest cells migrate to various positions in the body, where they become the pigment cells or chromatophores.

The Endodermal Derivatives

Basically, the endoderm gives rise to those structures associated with the primitive digestive tube. For convenience this primitive tube can be subdivided into the foregut, midgut, and hindgut. The endoderm of the foregut differentiates anteriorly into the lining of the mouth. More posteriorly, it gives origin to the lining of the pharyngeal pouches. These pouches are very important in the development of certain endocrine glands. The thymus glands develop from cells which bud off from the dorsal surface of the first and second visceral pouches (Fig. 9-39). The carotid glands, which are associated with the regulation of blood flow, develop from the ventral portion of the second visceral pouches. The third and fourth visceral pouches give rise to the parathyroid glands. The ultimobranchial bodies develop from the fifth visceral pouch. In the floor of the pharynx, between the left and right second visceral arches, an evagination develops, which later bifurcates and separates from the pharynx proper. This is the primordium of the thyroid gland.

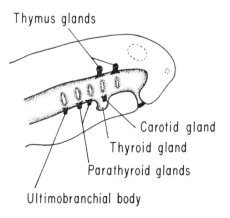

Figure 9-39. Diagrammatic representation of the pharyngeal region showing the more important derivatives.

The Respiratory System

Posterior to the last pharyngeal pouch, a midventral evagination occurs. This evagination is the laryngotracheal groove, the forerunner of the complete respiratory system. The opening into this evagination is the glottis. The distal end of the laryngotracheal evagination soon bifurcates. The unpaired region of the entire structure will develop into the larynx and trachea, whereas the bifurcated portions will become the bronchi and lungs of the adult.

Other Foregut Derivatives

The foregut posterior to the pharyngeal region becomes the esophagus. This is a narrow tube which connects the pharynx to the stomach. In cross section, the posterior end of the esophagus is indicated by an enlarged diameter of the gut wall as the saccular stomach is approached. Just posterior to the stomach, the liver arises as a ventral evagination which grows toward the anterior end of the coelomic cavity. This liver evagination gives rise, also, to the gallbladder. The original connection of the liver and gallbladder to the gut is called the hepatic or bile duct. At approximately the same region of the gut, another diverticulum, the pancreatic diverticulum, begins to grow. This develops into the pancreas and the pancreatic duct of the adult. The complete foregut, with the various formed structures, can be seen in Figure 9-40.

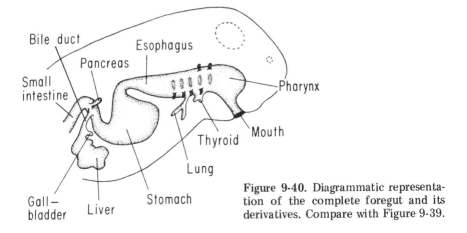

Figure 9-40. Diagrammatic representation of the complete foregut and its derivatives. Compare with Figure 9-39.

Midgut and Hindgut

The major portion of the development of the gut takes place in the foregut region. The midgut, which originally contained the large quantity of stored yolk, gradually decreases in diameter as the yolk is absorbed. Once the yolk absorption is completed, the midgut increases in length. This is consistent with the physiological requirements of a herbivorous animal. This midgut region, now called the small intestine, grows to a length which is many times that of the entire body. The hindgut region undergoes very little development beyond the rupture of the cloacal plate. During metamorphosis the entire digestive tube shortens considerably. This is related to the change in the dietary habits of the frog. As an adult, it is essentially carnivorous.

Mesoderm Development

When the mesoderm was last discussed, it was in three distinct, potentially different, portions: a medially located somite, an intermediate mesoderm, and a lateral plate mesoderm which was subdivided into an inner splanchnic and an outer somatic mesoderm. It is from this stage that the description of later mesodermal development begins.

Somites

During the course of the tadpole's development, the number of somites continues to increase in a posterior direction, until 40 or more pairs develop. Most of these, however, are lost when the tail is resorbed during the process of metamorphosis. Only 11 pairs remain to contribute to the formation of the definitive body structures. Each somite is composed of three distinct cellular areas and an inner cavity. The outer layer is the dermatome. Most of the dermal layer of the skin and the intermuscular connective tissues of both somatic and appendicular muscles are derived from the dermatomal cells. Separating the dermatome from the next more medial division of the somite is a cavity, the myocoele. This is a transitory structure which does not persist after the formation of the dermis and the further differentiation of the myotome. The myotome gives rise to the majority of the striated musculature of the body. Medial to the myotome is a group of somewhat less-organized cells termed, collectively, the sclerotome. These cells bud off from their original position in the somite and migrate to a site near both the neural tube and notochord. Here they form a sheet of cells, which soon surrounds these two structures. Figure 9-41 illustrates diagrammatically the development of a few typical vertebrae from these sclerotomal cells.

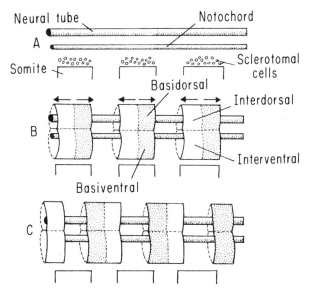

Figure 9-41. Vertebrae development.

Since the somites are arranged metamerically, it can be readily understood that the sclerotomal sheath is also segmented. Each segment is derived from the sclerotomal cells of a particular somite. The condensation of cells within this sclerotomal sheath is, however, greater in the posterior region of each segment. This allows us to distinguish four separate regions of the basic sclerotomal sheath. These are referred to as arcualia. Dorsally, there are a posterior basidorsal and an anterior interdorsal; ventrally, there are a posterior basiventral and an anterior interventral. Within a short period of time, the basidorsal and basiventral separate from the interdorsal and interventral. Each new segment now begins to migrate from its former partner. This migration results in the basidorsal and basiventral from each somite fusing with the interdorsal and interventral from the next most posterior somite. These four arcualia are the basic elements involved in the development of the vertebrae. From the basidorsal, the neural arch, centrum, and neural spine develop. The basiventral contributes to the hemal arch, hemal spine, and centrum. The interdorsal and interventral are lost, for the most part, in subsequent development, although part of their cellular structure must contribute to the centrum. This vertebral development is pictured in Figure 9-42.

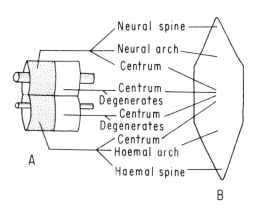

Figure 9-42. Adult derivatives of the basic arcualia. (A) Side view showing the basic arcualia formed from the sclerotomal cells. (B) Cross section through a generalized vertebra showing the structures derived from each arcualia.

The Mesomere

The mesomere or intermediate mesoderm is primarily responsible for contributing to the formation of the organs and the ducts of the excretory system. In the early frog larva this excretory system is primarily the pronephric or head kidney. Later in development this is replaced by a mesonephric kidney.

The Pronephric Kidney

In the anterior portion of the mesomere, a cavity is formed which soon becomes tubular. These tubular cavities of adjacent parts of the mesomeres soon join one another to form a continuous tube originating anteriorly. The tube, the pronephric duct, runs posteriorly to the cloaca. In the region of somites 2, 3, and 4, a second set of tubules begins to form in the mesomere. These, however, do not grow toward one another but toward the lumen of the pronephric duct in its own mesomere. The other end opens directly into the coelomic cavity. These small tubules, each completely contained within each segment, are the pronephric tubules. This development of both pronephric tubules and the pronephric duct results in an enlarged mass of tissue which is visible externally as the head kidney.

This kidney is never functional in the larval tadpole, although an abortive attempt is made to supply this kidney with a vascular system. Shortly after it is completely formed, it begins to degenerate and is replaced by the functional amphibian kidney, the mesonephros. The pronephros is seen in Figure 9-43.

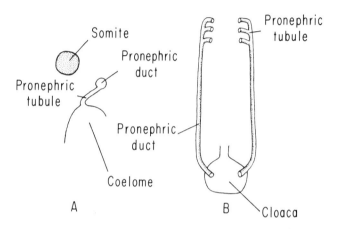

Figure 9-43. Diagrammatic representation of the pronephros as it would appear in cross section (A) and in gross dissection (B).

The Mesonephric Kidney

The mesonephric kidney (the mesonephros) develops posterior to the pronephric kidney. Usually, it occurs from the level of somites 7 through 12. The pronephric duct, which was formed in relation to the pronephric kidney, is called at this level the mesonephric duct. Segmented tubules, comparable to the pronephric tubules but more complex in their structure, soon develop within the substance of the intermediate mesoderm. These lead directly into the mesonephric duct. They become associated with the posterior cardinal vein and become functional by developing structures that enable them to extract waste products from the circulating blood in the posterior cardinal vein (Fig. 9-44). This mesonephric kidney continues to develop and enlarge throughout larval life and, in the adult, becomes the frog kidney.

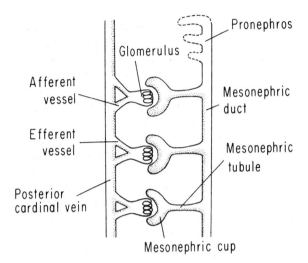

Figure 9-44. Diagrammatic representation of the mesonephric kidney, showing its relationship to both the pronephros and the posterior cardinal vein.

The Hypomere

When last described, the lateral plate mesoderm had separated into an inner splanchnic and an outer somatic layer. The space between these two layers is the coelomic cavity. In the frog, where no diaphragm develops, the coelomic cavity is known as the pleuroperitoneal cavity. One segment of this original coelomic cavity does become isolated from the rest. This is the pericardial cavity, which surrounds the developing heart.

Heart Development

In the region beneath the pharynx the splanchnic mesoderm comes together and forms a hollow tube. Within this tube, isolated cells, the endothelial cells, begin to organize themselves into an endothelial tube which grows both anteriorly and posteriorly. Posteriorly, this tube is double. Each part is called an omphalomesenteric vein. These omphalomesenteric veins grow directly to the yolk. Anteriorly, the original tube is also doubled. Here they are called the ventral aortae.

The unpaired endothelial tube becomes the endothelial lining of the heart, whereas the splanchnic mesoderm surrounding it becomes much thicker and forms the heart muscle or myocardium. The original coelomic cavity between this myocardium and the somatic mesoderm is, in this region, called the pericardial sac. The heart is attached both anteriorly and posteriorly. It is this attachment that prevents the heart from growing in these latter directions. The cells of the heart continue to reproduce mitotically at a very fast rate and, consequently, the tube increases in length. As the anterior and posterior attachment of the heart prevent it from expanding in this direction, the increase in the size of the heart is accomplished by a folding of the original tube into an S-shape (Fig. 9-45). Once this S-shape has been produced, certain regions of the tube can be described in terms of their adult nomenclature. Most posteriorly, where the two omphalomesenteric veins join, is a common tube, the sinus venosus. Just anterior to the sinus venosus is a thin-walled sac, the atrium. The tube now bends toward the ventral aspect of the body and becomes the large, thick-walled ventricle. This is followed anteriorly by the conus arteriosus and the bulbous arteriosus. This bulbous arteriosus bifurcates into two vessels, each of which is a truncus arteriosus. A

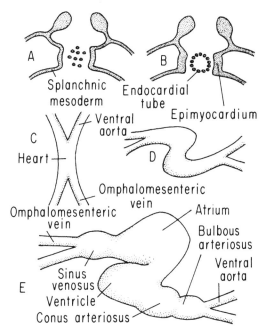

Figure 9-45. Development of the amphibian heart.

148   *The Embryology of the Frog*

further development of the heart causes a bifurcation of the original single atrium into a left and a right atrium. The adult frog heart is a three-chambered heart, composed of two atria and a single ventricle. The complete development of the amphibian heart, from the endothelial cells and splanchnic mesoderm, is depicted in Figure 9-45.

Development of the Vascular System

The Arteries

Originally, six aortic arches develop within the visceral arches. These, for the sake of clarity, can be depicted diagrammatically as in Figure 9-46. Connecting the ventral ends of these aortic arches to the truncus arteriosus of the heart are the paired ventral aortae. Dorsally, the aortic arches are connected to one another and to the more posterior regions of the body by the paired dorsal aortae. Further

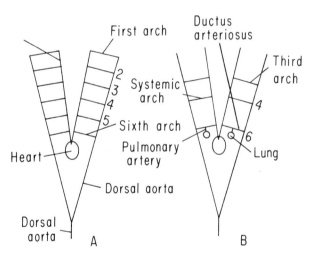

Figure 9-46. Early differentiation of the aortic arches. (A) The basic vertebrate aortic arch pattern includes the ventral and dorsal aortae and six aortic arches. (B) The first, second, and fifth arches are lost. A pulmonary artery develops from the sixth. That portion of the sixth arch between the pulmonary artery and the dorsal aorta is the ductus arteriosus.

posteriorly, these dorsal aortae fuse to form a single dorsal aorta. All these blood vessels form from isolated mesenchymal cells, which round up to form tubes. These isolated tubes ultimately coalesce to form continuous vessels. The first two aortic arches degenerate. The anterior extension of the ventral aorta then becomes known as the external carotid. The anterior extension of the dorsal aorta is the internal carotid. The next portion of the original aortic arch structure to degenerate is the dorsal aorta between the original third and fourth aortic arches. The fifth aortic arch also degenerates. This leaves the configuration shown in Figure 9-46. The original fourth arch becomes the major aortic arch of the adult frog. The sixth arch becomes the pulmonary arch, which will carry blood to the lungs. The completed aortic arch structure is represented both diagrammatically and as it actually occurs in Figure 9-47. As the dorsal aorta courses through the body, it gives off both paired and unpaired vessels. The paired vessels usually go to paired structures such as the arms, legs, kidneys, reproductive glands, and so on, whereas the unpaired vessels usually arise ventrally to be distributed to the visceral organs in the coelomic cavity. In the tail region, the dorsal aorta is called the caudal artery. This section of the dorsal aorta is lost as the metamorphic regression of the tail occurs.

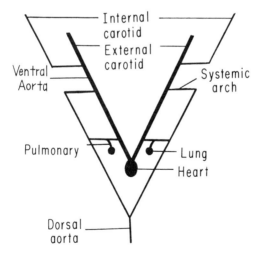

Figure 9-47. Later development of the aortic arches. The diagram illustrates the condition of the aortic arches after the first, second, and fifth aortic arches, as well as a portion of the dorsal aorta between the third and fourth arches, degenerate.

The Cardinal System

The primitive system of blood return in the embryonic frog is through the cardinal system. The paired anterior cardinals direct blood from the anterior areas of the body toward the heart, whereas blood is returned from the posterior regions of the body through the paired posterior cardinals (Fig. 9-48). These arise in the same manner as the aorta, that is, through the coalescence of various mesenchymal cells. In the region of the heart, the anterior and posterior cardinal veins of each side of the body fuse to form common cardinal veins, also called the ducts of Cuvier. These open directly into the sinus venosus of the heart. This primitive cardinal system undergoes extensive changes during the larval state of the frog. The anterior cardinal vein ultimately becomes known as the superior vena cava.

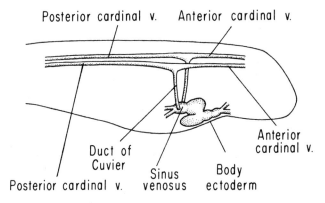

Figure 9-48. The embryonic cardinal vein system. Blood is returned from the anterior and posterior regions of the body by the anterior and posterior cardinal veins, respectively.

The Development of the Posterior Vena Cava

The posterior cardinal veins drain the blood from the posterior portion of the body into the common cardinal vein or duct of Cuvier. They lie lateral to the developing pronephros and mesonephros kidneys. Toward the medial surface of these kidneys, a new vessel, the subcardinal vein, develops (Fig. 9-49). The two subcardinal veins are pushed toward one another in the midline as a result of the growth of the kidneys. These eventually fuse to form the subcardinal sinus. Blood from the subcardinal sinus passes through the sinusoids of the mesonephros and is then carried directly to the posterior cardinal vein. The subcardinal sinus continues to grow anteriorly, until it fuses with the posterior cardinal in a region just posterior to its entry into the duct of Cuvier. Following this, that portion of the posterior cardinal vein between the entrance of the subcardinal and the mesonephros degenerates. This prevents blood from following the direct route to the heart and necessitates the flow of blood from the posterior regions of the body through the kidney and into the subcardinal sinus before it returns to the heart. This is the embryonic renal portal system, similar to that found in the adult fish forms.

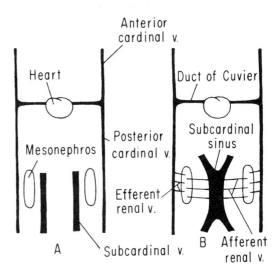

Figure 9-49. The development of the subcardinal sinus. (A) A new vein, the subcardinal vein, develops ventromedial to each mesonephric kidney. (B) As a result of kidney growth, the subcardinal veins are pushed closer together until they coalesce, forming the subcardinal sinus.

A new vessel, the posterior vena cava, begins to bud off from the duct of Cuvier at a point near the heart. This new vessel grows caudally until it fuses with the subcardinal sinus. After this fusion, the anterior portion of the subcardinal sinus (that portion between the duct of Cuvier and the sinus itself) degenerates. The resulting vessel, made up of the posterior portion of the subcardinal sinus and the new postcaval vein, is now known throughout its length as the postcaval or posterior vena cava. As this continues to grow caudally, the posterior cardinal vein lateral to the kidney begins to degenerate. The major vein, now draining the posterior regions of the body, is the posterior vena cava. The development of this posterior vena cava is shown in Figure 9-50.

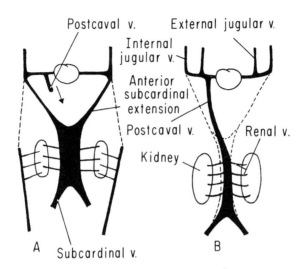

Figure 9-50. Development of the postcaval vein. (A) Anterior extensions of the subcardinal vein grow toward and fuse with the posterior cardinal vein just posterior to its entry into the duct of Cuvier. That portion of the posterior cardinal represented by the dotted lines then degenerates. Blood returning from the posterior regions of the body must now pass through the kidney on its way to the heart. A new vein, the postcaval vein, begins to grow posteriorly from the duct of Cuvier. (B) This postcaval vein ultimately fuses with the subcardinal sinus. All the veins represented by dotted lines then degenerate. The original anterior cardinal vein is renamed the internal jugular vein. A new vein forms in the floor of the mouth. This is the external jugular vein (see also Fig. 9-49).

Hepatic Portal System

As the liver primordium grows ventrally, it meets and grows around the omphalomesenteric or vitelline veins. That portion of the veins surrounded by the liver substance soon degenerates into a complex network of sinusoids. The veins now located between the liver and the heart are the hepatic veins, whereas the portions found between the midgut and the liver are the hepatic portal veins (Fig. 9-51).

The double hepatic portal vein now anastamoses along its entire length. Subsequent degeneration of segments reduces this original double system to a single hepatic portal vein (Fig. 9-52). With the completion of the hepatic portal system, the circulatory pattern appears as in Figure 9-53.

Figure 9-51. Development of the hepatic portal system. (A) As the liver diverticulum continues to grow, it encounters the omphalomesenteric veins which connect the yolk with the sinus venosus. (B) As the liver grows around the omphalomesenteric veins, they break up into sinusoids. The portion of the original vein between the liver and the heart becomes the hepatic vein, whereas that portion between the yolk and the liver is the hepatic portal vein (see also Fig. 9-52).

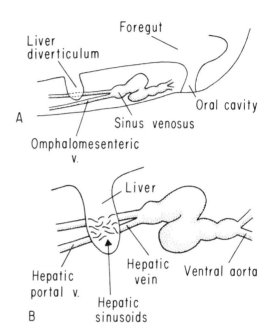

Figure 9-52. The omphalomesenteric veins anastomose along their entire length. Subsequent degenerations (dotted lines) result in a single hepatic portal vein connecting the duodenum to the liver.

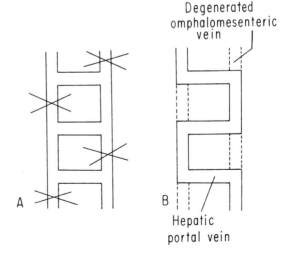

Figure 9-53. Completed circulatory system of the metamorphosed frog.

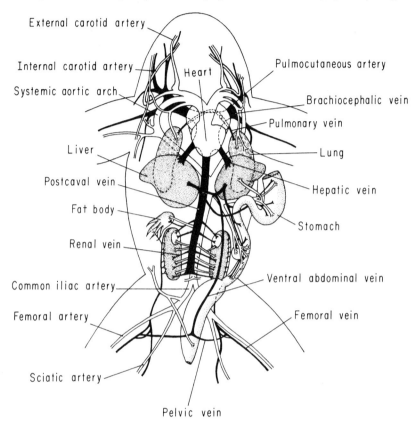

Reproductive System

In both sexes, a middorsal thickening of the gut, the sex cell ridge, is the first indication of the developing gonads. This thickening soon separates from the gut and is then known as the sex cell cord. This sex cell cord then splits longitudinally into two genital ridges. The cells of the genital ridges are the future spermatogonia and oögonia of the adult gonads. These genital ridges become surrounded and infiltrated by the neighboring cells of the mesomere and peritoneum, and differentiate into the ovaries and testes.

The mesonephric tubules are used by the testes as the vas efferens, while the mesonephric duct becomes the wolffian or sperm duct (also called the vas deferens). The mesonephric duct in the male, therefore, serves as a true urogenital duct, carrying both urine and sperm.

In the female, a müllerian duct develops from the cells of the degenerating pronephros. This duct serves only a reproductive function. A similar duct develops in the male, but it remains vestigial.

"I have taught thee in the way of wisdom; I have led thee in right paths. When thou goest, thy steps shall not be straitened; and when thou runnest, thou shalt not stumble. Take fast hold of instruction; let her not go; keep her; for she is thy life." Proverbs 4:11-13.

Chapter 10

The Embryology of the Reptile

In all the specific forms studied thus far, the Amphioxus, the Teleost, and the Amphibian, the eggs have been deposited directly into the water, and fertilization has been external. This statement is in general, correct, although ovoviviparous exceptions have been noted. This type of reproductive pattern, i.e., where the environmental water is the site of fertilization, is efficient. It does, however, have one major drawback. It requires that either the animals live in an aquatic environment or that they return to water for their spawning period. In a way, this chains these animals to a complete, or at best partially, aquatic environment.

The reptiles, birds, and mammals are not faced with this problem for they develop from a different egg type. This egg, the cleidoic egg, is unique in that the embryo is surrounded by a membranous sac which contains water. The entire egg is covered with a semipermeable shell which prevents dessication. Other membranes hold the stored food (yolk) while still others collect the embryo's metabolic wastes. This egg, therefore, frees the animal from its dependence on environmental water as a site for development. Reptilian eggs can be laid on arid, sandy soils and develop normally. In most respects, the egg is similar to that of the chick. A complete discussion of this egg type is, therefore, deferred until Chapter 11, "The Embryology of the Chick." The student is referred to that chapter for a description of both the morphology and functional significance of this type of egg and the extraembryonic membranes which were not present in the noncleidoic eggs studied thus far.

Egg Morphology

The reptile selected for this study is the painted turtle, *Chrysemys picta*. Since the turtle egg develops without an external source of food for a relatively long period of time, it contains a large quantity of stored nutritive material, the deutoplasm or yolk. This means that the egg is of the polylecithal type, and yolk is distributed according to the heavily telolecithal pattern. The egg is ovulated from the ovary when it is in the metaphase II stage of meiosis. Fertilization in this form is internal. The sperm are deposited in the cloaca during the copulatory act. These sperm then actively swim from their point of deposition in the cloaca, through the vagina, uterus, and oviduct, until they reach the osteum, the opening into the oviduct. Here they meet the ovulated egg.

Fertilization

Polyspermy, or the entrance of more than one sperm into an egg, is the general rule in reptiles. It has been suggested that the underlying factor that accounts for the entrance of many sperm in the region of the germinal disk is that the presence of a large quantity of yolk inhibits the normal cortical reaction. Although many sperm enter the egg, only one is functional. The rest degenerate. This functional sperm or male pronucleus, as it is now called, soon fuses with the female pronucleus.

Although the fertilization membrane lifts from the surface of the egg, creating a perivitelline space in all vertebrates, the rate at which this is accomplished differs. In the turtles it is much slower than in the amphioxus and teleost fish. Once this fertilization membrane lifts, the egg is covered with albumen (egg white) in the albumen-secreting portion of the oviduct. This albumen layer is then covered with an egg shell of a brittle or semibrittle nature. The color of reptilian eggs is always white or off-white. It is never highly colored, as is the case in many bird eggs.

Once the fertilized egg is covered with the albumen and shell, it is laid. The turtle digs a nest, with her hind feet, in a sunny spot along the water's edge. The eggs are laid in this carefully prepared hole and are then covered with a layer of dirt. After laying her eggs, the mother turtle takes no further interest in the development of either the eggs or the newly hatched turtles. The incubation time, under these conditions, will vary because it is dependent on the temperature to which the developing eggs are subjected.

Cleavage

As the turtle egg is a heavily telolecithal type, cleavage is confined to the germinal disk. This type of cleavage has already been classed as meroblastic or discoidal. The first cleavage plane occurs across the short axis of the germinal disk. This separates the original uncleaved germinal disk into two equal blastomeres. The second cleavage plane is at right angles to the first. This results in four equal blastomeres. The third cleavage plane, oriented at right angles to plane two and parallel to plane one, produces two rows of four blastomeres each. The fourth cleavage plane is parallel to plane two and at right angles to planes one and three. This produces four rows of four cells each. Future cleavages are irregular. The marginal cells are continuous with the deutoplasm. The central cells are completely delimited by cellular membranes. Figure 10-1 is a diagrammatic representation of the cleavage disk after the fourth cleavage plane has occurred.

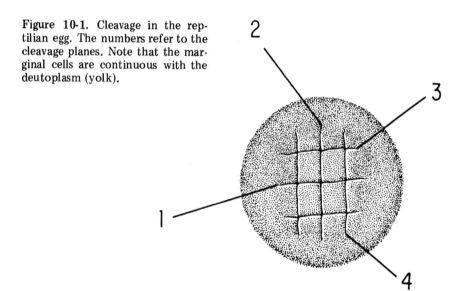

Figure 10-1. Cleavage in the reptilian egg. The numbers refer to the cleavage planes. Note that the marginal cells are continuous with the deutoplasm (yolk).

Formation of the Periblast

As a result of the cleavage process, a mound of cells, the blastoderm, soon occupies the space originally held by the uncleaved germinal disk. Some of the marginal cells begin to migrate between this blastoderm and the yolk. Other blastodermal cells migrate away from the blastoderm. All these cells that leave the blastoderm are called periblast cells. They soon lose their cellular identity and form a syncytial mass beneath and to each side of the original blastoderm. That portion which is beneath the blastoderm is referred to as the central periblast, whereas the periblast surrounding the margins of the blastoderm is the peripheral periblast. This peripheral periblast is also called the germ wall. The original blastoderm is composed of embryonic tissue which will develop into the embryo itself. The space between the blastoderm and the central periblast is called the primary blastocoele. A sagittal section through the early blastula is seen in Figure 10-2.

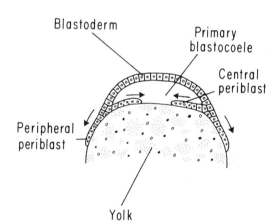

Figure 10-2. Sagittal section through an early reptilian blastula.

Blastulation

As a result of the movements described, a primary blastula is formed. A sagittal section through this primary blastula is seen in Figure 10-3. A dorsal view (Fig. 10-4) reveals that the central blastoderm, being lifted from the surface of the yolk, is optically less dense than the marginal blastoderm. This central blastoderm area is frequently referred to as the area pellucida. The area pellucida has beneath it the primary blastocoele. The surrounding edge, that which

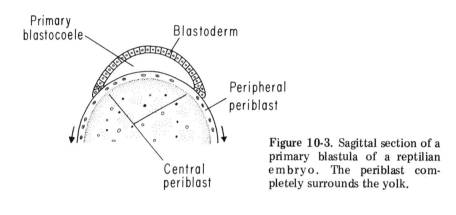

Figure 10-3. Sagittal section of a primary blastula of a reptilian embryo. The periblast completely surrounds the yolk.

Figure 10-4. Dorsal view of the primary blastula depicted in Figure 10-3. The area pellucida appears lighter because it is separated from the underlying yolk by the primary blastocoele.

is continuous with the yolk, is called the area opaca. This is also continuous with the syncytial peripheral periblast. A secondary blastula next forms. The original blastoderm produces, as a result of horizontal cleavages, two layers, an upper epiblast and a lower hypoblast. The formation of this hypoblast has led to three general theories regarding its production. Some authorities believe that cells leave the original blastoderm and infiltrate to the surface of the periblast. This process is called infiltration. Others believe that the original blastoderm splits into two layers, the process known as delamination. Still others think that part of the blastoderm turns in at its edge to grow inward over the yolk. This is the process of involution. Regardless of which process ultimately is discovered to be the true process, a second layer is formed. The upper layer, the epiblast, contains the materials for the presumptive epidermis, mesoderm, neural tissue, and notochord. In addition, a limited amount of presumptive endoderm is present. The hypoblast is the presumptive endoderm. The space between the epiblast and hypoblast is the secondary blastocoele. Both the hypoblast and epiblast are connected peripherally with the peripheral periblast. The central periblast soon degenerates, allowing the hypoblast to come into intimate contact with the yolk beneath. A fate map of the turtle blastula is seen in Figure 10-5.

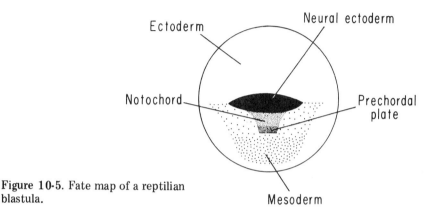

Figure 10-5. Fate map of a reptilian blastula.

Histologically, the epiblast differs from the hypoblast. The upper layer or epiblast is composed of flattened cells which are closely packed. The hypoblast consists of rounded cells which are loosely arranged. The spaces between the hypoblast cells are evidently filled with a fluid-type medium. In the posterior region of the area pellucida, a thickened area which is slightly pear-shaped begins to appear. This thickened region is the embryonic shield. Its optical density is due to the differentiation of the epiblast cells in the area from a squamous to a columnar type. Immediately posterior to the embryonic shield is a primitive plate (Fig. 10-6). It is in this region of the primitive plate that the major gastrulation movements occur.

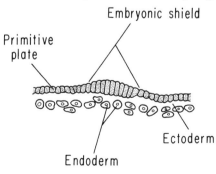

Figure 10-6. Sagittal section through the blastoderm showing the formation of the embryonic shield. The primitive plate is found just posterior to the embryonic shield.

Gastrulation

A depression soon occurs in the primitive plate region. This is homologous to the dorsal lip of the blastopore in the amphibians and, as will be seen, homologous also to the primitive pit of the avian embryo. This indentation forms a cavity, the archenteron. This cavity has been given various names by investigators. Among these names are the blastoporal canal and notochordal canal. I prefer to use the term "archenteric canal." This cavity first forms vertical to the epiblast. Soon, however, the epiblast begins to shift posteriorly,

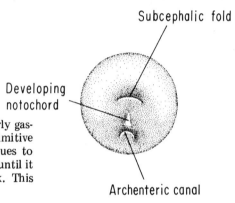

Figure 10-7. Dorsal view of the early gastrula. An invagination of the primitive plate region develops. This continues to grow both anteriorly and ventrally until it finally breaks through to the yolk. This canal is the archenteric canal.

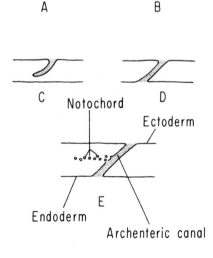

Figure 10-8. Development of the archenteric canal and notochord. (A) Epiblast and hypoblast. (B) An early view of the invagination of the archenteric canal. (C and D) This canal continues to grow both anteriorly and ventrally until it meets and breaks through the hypoblast. (E) The notochord arises from cells proliferating from the roof of this archenteric canal. The epiblast is now called ectoderm, and the hypoblast is referred to as endoderm.

creating an oblique condition of the archenteric canal. The opening itself, at first round, begins to take on the appearance of an inverted U (Fig. 10-7). This archenteric canal should be likened to the archenteron of the frog. The roof, i.e., the portion most anterior, is truly mesendoderm and will give rise to the notochord, mesoderm, and roof of the gut. As the archenteric cavity continues to grow inward, it ultimately fuses with the hypoblast and then breaks through, opening directly to the yolk. This continuous tube, running from the epiblast to the hypoblast, persists for most of the development of the turtle. Figure 10-8 shows the development of the archenteric canal and the development of the notochord and mesoderm. Figure 10-9 is a cross section through a postgastrular turtle embryo anterior to the archenteric canal.

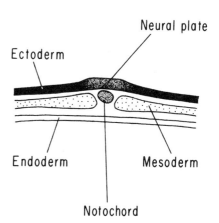

Figure 10-9. Cross section through a postgastrular turtle embryo anterior to the archenteric canal.

Neurulation

Neurulation in the reptilian embryo is, in certain respects, very similar to that in the frog. The thickened neural plate, now underlain by the notochord, begins to thicken appreciably and round up at its lateral edges. These up-pushings, the neural folds, begin to approach one another toward the midline. The closure of the neural folds does not occur at the same time throughout the anterior-posterior axis. The closure occurs first a short distance anterior to the external opening of the archenteric canal. From this position the folds close both anteriorly and posteriorly, much as a zipper would close a seam. Figure 10-10 depicts an embryo during the process of neurulation. The last portion of the neural tube to close anteriorly is referred to as the anterior neuropore. Posteriorly, the folds close over the archenteric canal. In reality, this archenteric canal connects the neurocoele with the gut. Figure 10-11 is a sagittal section of a turtle embryo after the formation of the neural tube.

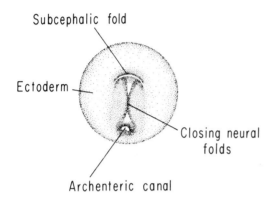

Figure 10-10. Turtle embryo during the closure of the neural folds. The closure occurs first, a short distance anterior to the external opening at the archenteric canal. The folds then proceed to close in both an anterior and posterior direction.

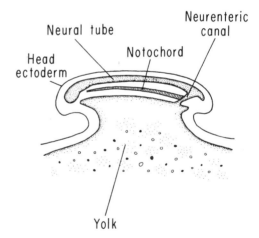

Figure 10-11. Sagittal section of a turtle embryo after the formation of the neural tube. Note that a connection exists between the neural tube and the gut. This is the neurenteric canal.

The Formation of Body Shape

The formation of body shape is very similar to that already seen in the teleost embryo. The three primary germ layers, as well as the notochord and neural tube, are now lying flat on the yolk. An undercutting occurs first in the head region. This undercutting is the subcephalic fold. This, in reality, elevates the head area from the rest of the blastoderm. This fold itself contains elements of ectoderm, endoderm, and mesoderm. It creates, for the first time, an anterior gut portion, which is now completely lined with endoderm. Figure 10-12 is a sagittal section through the anterior portion of the turtle embryo after the formation of this subcephalic fold. Once the fold has been established, an anterior growing process is initiated which produces a

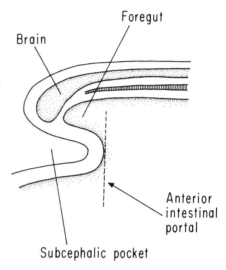

Figure 10-12. Sagittal section through the anterior end of the turtle embryo after the formation of the subcephalic fold.

definitive head structure. As the head continues to develop, two modifications of the growing process occur. One is a process referred to as flexion, in which the embryo bends at strategic points along its anterior-posterior axis. These points are the midbrain region, the cephalic flexure; the cervical region, the cervical flexure; and the tail region, the caudal flexure. Another process is torsion. This entails a rotation of the body on its anterior-posterior axis so that the embryo comes to lie on its left side on the yolk.

In the tail region, a fold, similar to that seen in the head, also occurs. This subcaudal fold delimits the tail from the rest of the blastoderm. A section through the tail, showing the germ layers after the initiation of the subcaudal fold, can be seen in Figure 10-13. A complete embryo, after the production of the head and tail regions and after flexion and torsion have occurred, is seen in Figure 10-14.

The folds that established the head and tail also occur in the body region. These lateral folds also lift the embryo's body from the

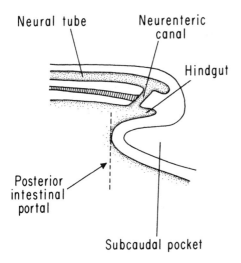

Figure 10-13. Sagittal section through the developing tail region after the formation of the subcaudal fold.

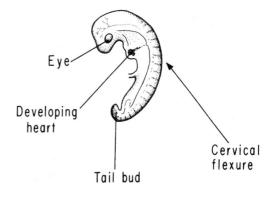

Figure 10-14. A turtle embryo, after the production of the head and tail regions, and after flexion and torsion have occurred.

yolk and serve to delimit the embryonic body from the extraembryonic membranes. These various folds help to delimit the three primary areas of the gut. That portion of the digestive tube which extends anterior to the open yolk is now termed the foregut. The opening into the foregut is the anterior intestinal portal. That portion which extends posterior to the open yolk area is the hindgut. The opening into the hindgut is the posterior intestinal portal. The most extensive portion, that which lies open to the yolk, is the midgut. Figure 10-15 illustrates diagrammatically these three primary gut areas.

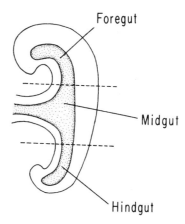

Figure 10-15. Diagrammatic representation of the turtle, delimiting the three primary gut regions which result from the subcephalic and subcaudal foldings.

Later Development of the Turtle

The later development of the turtle, that is, the formation of the primordia of the various organ systems, is very similar to that of the chick, and discussion of it will be left for Chapter 11 to avoid unnecessary repetition. One aspect of turtle development is, however, different enough to warrant discussion here. The turtle possesses a dermal skeleton in the form of a dorsal carapace and a ventral plastron. These develop from the mesenchymal cells found in the dorsal and ventral region of the embryo. They do not occur until relatively late in development, but once they make their appearance, they give the embryo the characteristic look of a turtle embryo. Once the carapace is formed, the embryonic turtle appears as in Figure 10-16.

One other aspect of reptilian development is different enough to warrant a separate discussion. This is the modification of the heart

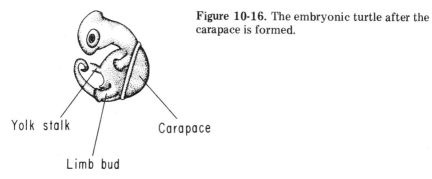

Figure 10-16. The embryonic turtle after the carapace is formed.

development. In the reptile we perceive an intermediate form between the three-chambered heart of the frog and the four-chambered heart of the birds and mammals. This intermediary step is represented by a partial interventricular septum, which develops after the formation of the single ventricle. In reality then, the reptile has a four-chambered heart, although the two portions of the ventricle are not completely separated. In one group of reptiles, the crocodilia, this separation is complete, and a true four-chambered heart develops. The rest of the organ systems and their embryonic development are covered in the next chapter.

"Behold the fowls of the air: for they sow not, neither do they reap, nor gather into barns; yet your heavenly Father feedeth them. Are ye not much better than they?"
Matthew 6:26.

Chapter **11**

The Embryology of the Chick

The development of the organ systems of the reptile, bird, and mammal are similar enough to warrant a general discussion for all three forms rather than to repeat the organogenesis for each of the different embryos. It will be the purpose of this chapter, therefore, to discuss not only the basic embryological patterns of cleavage, blastulation, gastrulation, and neurulation, but also to discuss the primary organ systems of the bird and indicate their similarity to both the reptilian and mammalian embryos. It will then be sufficient in the chapter on mammalian embryology, as it was in the previous chapter on reptilian embryology, to discuss only the basic early embryonic stages.

The embryo chosen for this particular study is that of the common domestic fowl. This egg has been studied perhaps more intensely than any other vertebrate egg. There are many good textbooks concerned solely with the development of the chick. The student is referred to these for additional information on avian development.

Egg Morphology

The egg of the bird is another example of a cleidoic egg. This egg is designed to allow development to take place outside the maternal body. Since the period of development is approximately 21 days, the egg must contain within it enough yolk, or deutoplasm, to supply the embryo for the entire period of development. To prevent it from drying out, a hard shell that is relatively impervious to water is necessary. In addition, the embryo must be surrounded by a fluid. Surrounding this centrally located yolk is a vitelline membrane. Immediately outside this membrane is a considerable amount of watery material, the albumen or white of the egg. The outer calcium-impregnated shell is lined on the inside by two shell membranes. These may be called the outer and inner shell membranes. A sagittal section through a chick egg is seen in Figure 11-1. This egg, having a preponderance of yolk, is the polylecithal type, the yolk being distributed according to the heavily telolecithal pattern. In this respect it is similar to both the fish and reptilian eggs.

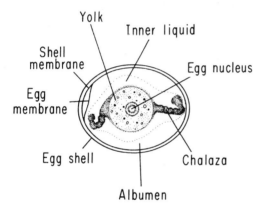

Figure 11-1. Sagittal section through a chick egg.

Fertilization

Fertilization in the chick occurs in the upper region of the oviduct. The passage of the fertilized egg is slow, taking approximately 22 to 24 hours. This means that most of the early development, i.e., cleavage and formation of the epiblast, takes place in the body of the hen before the egg is laid.

Cleavage

As a result of the heavily telolecithal yolk distribution, cleavage is limited to the germinal disk. This is cleavage of the meroblastic type or discoidal cleavage. The early cleavage planes are all vertical. The first cleavage plane begins at the animal pole and proceeds partly through the germinal disk. The second cleavage plane occurs at right angles to the first. It will be noted from Figure 11-2 that these cleavage furrows are incomplete, resulting in incomplete blastomeres. The cleavage planes after the second occur rather irregularly. They do have the effect, however, of separating the central cells from the more peripheral cells. The central cells have complete cell borders, whereas the marginal cells, or blastomeres, lie in only one plane as a result of the vertical cleavages.

The yolk in the central portion of the blastodisk is either utilized or the cells lift from the surface. In any event, the central cells are separated from the underlying yolk by a cavity, the primary blastocoele. This creates an optical differential when it is viewed under a light microscope. Figure 11-3 illustrates a dorsal view of a chick blastoderm after this primary blastocoelic cavity has formed. It will be noted that the central area, the area pellucida, is lighter than the surrounding area, the area opaca. The area opaca cells are still in direct contact with the underlying yolk.

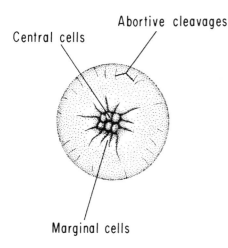

Figure 11-2. Early cleavage in the chick egg. Only the central cells have complete boundaries.

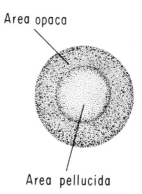

Figure 11-3. Dorsal view of the blastoderm after cleavage. The area pellucida appears lighter, for beneath it is the primary blastocoele.

Gastrulation

A second layer is eventually formed, which lies on top of the yolk. This layer, the hypoblast, originates from cells that have left the original blastoderm. Three primary hypotheses have been advanced to explain the origin of this layer. Some investigators believe that cells simply leave the area to drift down to the yolk. This is known as the method of infiltration. Others think that the cells of the blastoderm are delaminated, that is, horizontal cleavages occur which separate the original single layer of cells into two layers. A third school favors an involution hypothesis, indicating that cells in the peripheral area of the blastoderm turn under and grow over the surface of the yolk. In any event, a blastula is formed which, in a sagittal section, appears as in Figure 11-4. This blastula has an upper epiblast, a lower hypoblast, and a blastocoele cavity between the two. The hypoblast is found to lie directly on the yolk. It is at this stage that the egg is laid.

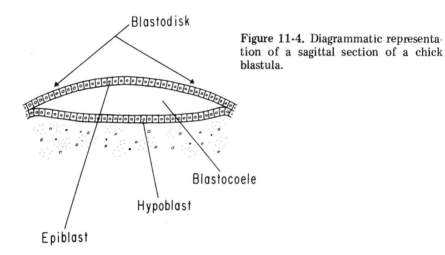

Figure 11-4. Diagrammatic representation of a sagittal section of a chick blastula.

The central region, the area pellucida, is the embryo-forming portion of the blastula. In the posterior region of the area pellucida a streak forms, which runs approximately two-thirds the distance from the posterior edge of the area pellucida. This primitive streak is a depression with raised areas on either side, the primitive folds. Anteriorly, the streak deepens to form a primitive pit. The raised area anterior to the primitive pit is the primitive, or Hensen's node. This stage is reached at approximately 20 hours of incubation. Figure 11-5 is a dorsal view of the 20-hour chick, showing the primitive streak.

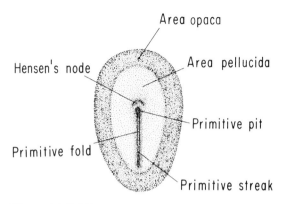

Figure 11-5. Whole mount of a 20-hour chick embryo, showing the location of the primitive streak.

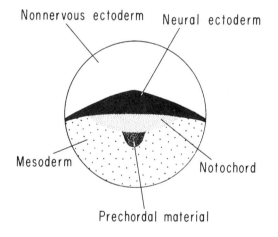

Figure 11-6. Fate map of a chick blastoderm. All the areas are at this stage, presumptive.

It is important to isolate the presumptive areas of the chick blastoderm in order to fully appreciate the gastrular movements involved in the chick egg. Figure 11-6 illustrates a fate map of the chick blastoderm. Investigators have marked distinct areas of the blastoderm with india ink in an attempt to observe the movement of these india ink particles during the gastrular process. Figure 11-7 shows the general direction of epiblast movement toward the primitive streak. These areas, moving toward the primitive streak,

Figure 11-7. Epiblast movement during chick gastrulation. The arrows indicate the direction of cellular movement toward the primitive streak.

involute around the primitive ridges and Hensen's node, to form a third germ layer between the hypoblast and epiblast. This third germ layer, composed of mesoderm and the notochord (also called the head process), can be seen forming in Figure 11-8. As this mass migration of epiblast cells occurs, the entire primitive streak begins to recede toward the posterior boundary of the area pellucida. This gives the mesoderm the appearance of two distinct anteriorly directed wings, and the notochord, of a single medial outgrowth. A

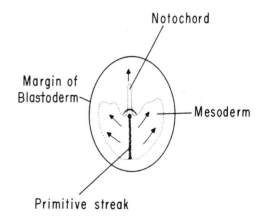

Figure 11-8. Formation of the mesoderm and notochord (head process). After the epiblast involutes around the primitive ridges and Hensen's node, it moves out between the epiblast and hypoblast in the direction indicated by the arrows.

cross section through the chick at this stage, just anterior to the Hensen's node, can be seen in Figure 11-9.

An interesting sidelight develops here. There is no true archenteron in the chick. In looking at the development of the frog and the reptile, we find that the cells that involute around the dorsal lip in the case of the frog and those that involute into the notochordal or archenteric canal in the reptile are similar to those that involute around Hensen's node in the chick. A small cavity does exist in the notochordal mass immediately anterior to Hensen's node. There is no other cavity in the chick that can be compared to the archenteron. The head process or notochord continues to develop anteriorly, delimiting the future embryonic body. The streak itself continues to shorten. Figure 11-10 shows the relative positions of the streak during the gastrulation process.

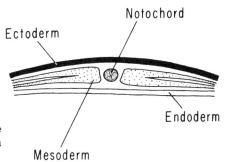

Figure 11-9. Cross section through the postgastrular chick embryo, through a region anterior to Hensen's node.

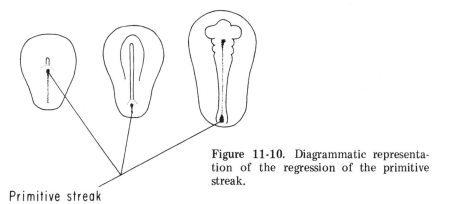

Primitive streak

Figure 11-10. Diagrammatic representation of the regression of the primitive streak.

Neurulation

The ectoderm immediately above the notochord thickens appreciably to form the medullary plate. This medullary plate then begins to round up at its edges into neural folds, which gradually elevate and grow toward one another in the midline. This process of neurulation, or neural tube formation, occurs most rapidly in the area some distance anterior to Hensen's node and proceeds in both anterior and posterior directions. Eventually, a neural tube forms throughout the embryo, except in the area of the primitive pit and streak, where a certain amount of involution is still occurring. The neural folds in this region fan out to create a rhomboidlike depression called the sinus rhomboidalis. When the neural tube has completely formed, the embryo appears as in Figure 11-11. The last region of the neural fold to close anteriorly is known as the anterior neuropore. The cavity of the neural tube is referred to as the neuro-

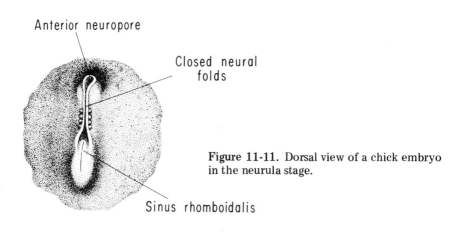

Figure 11-11. Dorsal view of a chick embryo in the neurula stage.

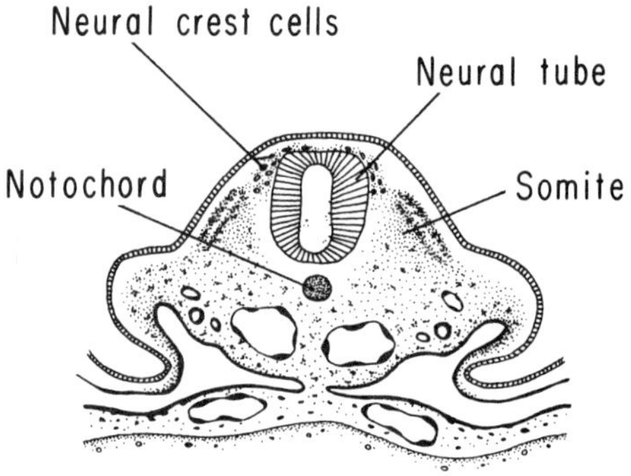

Figure 11-12. Cross section of a chick embryo of approximately 48 hours incubation, showing the completed neural tube and the neural crest cells.

coele. The formation of this neural tube results also in the formation of special cells which are located in the area of the fused neural folds. These are the neural crest cells, whose development will be followed later. These structures can be seen in Figure 11-12.

Anteriorly, the head area begins to lift from the blastoderm as the result of an undercutting subcephalic fold. This creates a subcephalic pocket which separates the ectoderm of the blastoderm from the ventral ectoderm of the head. The hypoblast, now properly called the endoderm, is likewise involved in this folding process. Prior to the establishment of this head fold, the entire endoderm was exposed to the yolk mass. As a result of the fold, the anterior portion of the endoderm is now a pocket. In reality, this anterior portion of the gut now has a roof and a floor, as compared with the rest of the endoderm which has no floor. On the basis of this distinction, the anterior portion is called the foregut, and the rest of the gut region, that is, that portion without a floor, is the midgut. The opening into the foregut is the anterior intestinal portal. Figure 11-13 shows a cross section through a region anterior to the anterior intestinal portal and another section immediately posterior to this point.

The involuted mesoderm begins to differentiate within a short period of time. That portion nearest the notochord remains thick-

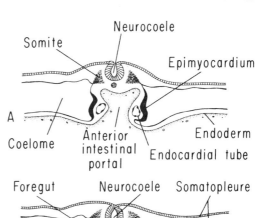

Figure 11-13. Anterior intestinal portal. As a result of the subcephalic fold, the endoderm in the head becomes a pocket, the foregut. The entrance to the foregut is the anterior intestinal portal. (A) Immediately posterior to the anterior intestinal portal. (B) Anterior to the portal.

ened. This is the epimere. Lateral to the epimere is a thin constricted portion, the mesomere or intermediate mesoderm. This will develop later into the excretory structures of the embryo. Lateral to the mesomere is the hypomere or lateral mesoderm. This splits into two distinct layers, an outer somatic mesoderm and an inner splanchnic mesoderm. The space between these two layers is the coelomic cavity. The somatic mesoderm and ectoderm are sometimes referred to as a single layer, the somatopleure, whereas the splanchnic mesoderm and endoderm are called collectively the splanchnopleure. A cross section of a 24-hour chick embryo (Fig. 11-14) reveals this mesodermal differentiation.

It is customary, when speaking of stages beyond this point, to refer to the hour of incubation. The most widely-used stages are the 33-hour, 48-hour, 72-hour, and 96-hour chick embryos. This format will be followed in the ensuing discussion. After each stage, the individual organ systems will be discussed in terms of their development to that point.

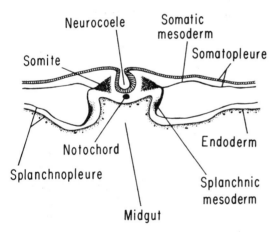

Figure 11-14. Cross section of a 24-hour chick embryo in the region of the heart. The salient points of mesodermal differentiation are depicted.

The 33-Hour Chick

Nervous System

At 33 hours of incubation, the neural tube has already closed throughout most of its length. A small opening persists until the 33-hour stage in the anterior end of the neural tube. This is the anterior neuropore. Posteriorly, the neural folds flare out around the primitive streak, creating a rhomboidlike depression, the sinus rhomboidalis.

Within the anterior portion of the neural tube an enlargement occurs that is to become the definitive brain. Within this bulbous enlargement, three brain regions can be distinguished. The most anterior is the prosencephalon. This is followed by the mesencephalon and, most posteriorly, by the rhombencephalon. In the region of the prosencephalon, the first of many differentiations can be seen from a dorsal view (Fig. 11-15). These are the optic vesicles, the forerunners of the eyes. In the floor of the prosencephalon an evagination,

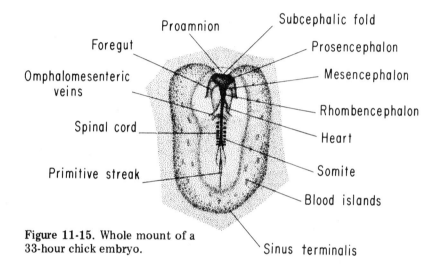

Figure 11-15. Whole mount of a 33-hour chick embryo.

directed ventrally, occurs. This is the infundibulum. This infundibulum will develop into the posterior portion of the pituitary gland. Figure 11-15 represents a dorsal view of a 33-hour chick embryo. Figure 11-16 illustrates selected cross sections of this embryo. In addition to the primary brain regions, a characteristic feature is the segmented somites in the posterior region. These somites represent segmentation of the epimere. This process of segmentation occurs in an anterior-posterior direction. Approximately one pair of somites is added each hour after 21 hours. Thus, the 33-hour chick has approximately 12 pairs of somites. No appreciable internal differentiation within the somites occurs up to 33 hours.

The foregut, which has already been formed as a result of the subcephalic folding, continues to lengthen as the backward movement of the subcephalic fold progresses. This moves the anterior intestinal portal some distance posterior.

180 The Embryology of the Chick

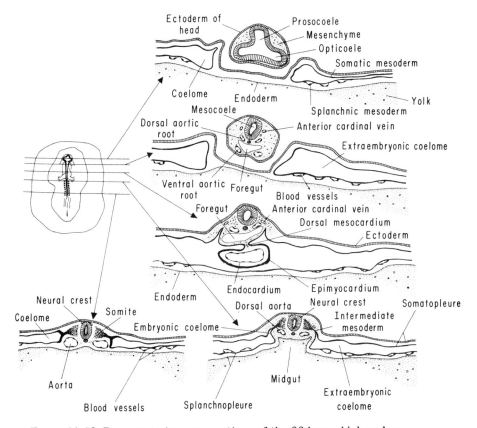

Figure 11-16. Representative cross sections of the 33-hour chick embryo.

A pair of conspicuous vessels develops within the splanchnopleure. These vessels, the omphalomesenteric veins, serve to connect the yolk mass with the developing heart. Their development will be followed at the 48-hour stage.

The 48-Hour Chick

The period between 33 and 48 hours of development is a very significant one for the embryo. A number of changes occur that create an embryo having a completely different appearance from that seen at the 33-hour stage. Figure 11-17 represents the whole mount of the 48-hour chick, and Figure 11-18 some representative cross

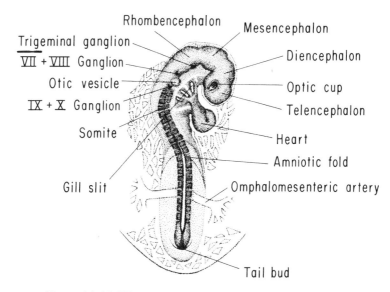

Figure 11-17. Whole mount of a 48-hour chick embryo.

sections. The most obvious change is in the overall appearance of the embryo. Instead of lying with its ventral surface in contact with the yolk, it is now partly turned so that the anterior portion of the body is lying with its left side to the yolk. This process, a twisting of the entire body along the anterior-posterior axis, is referred to as torsion. Another type of movement, which results in the new "bent" appearance, is flexure. In the head region the neural tube bends in the region of the mesencephalon. This is the cranial flexure. Another slight flexure occurs in the posterior region of the brain. This is the cervical flexure. These two flexures give the anterior portion of the embryo a C-like structure. It also raises problems when studying an embryo from serial sections. It can be seen that as one proceeds posteriorly, when studying a serially sectioned embryo, in the anterior portion of the neural tube the sections are actually proceeding anteriorly. It also means, as can be seen in Figure 11-18, that many sections will show two neural tubes, two notochords, two foreguts, etc. If the appearance of the whole mount of the 48-hour chick is referred to when serial sections are studied, this difficulty can be overcome.

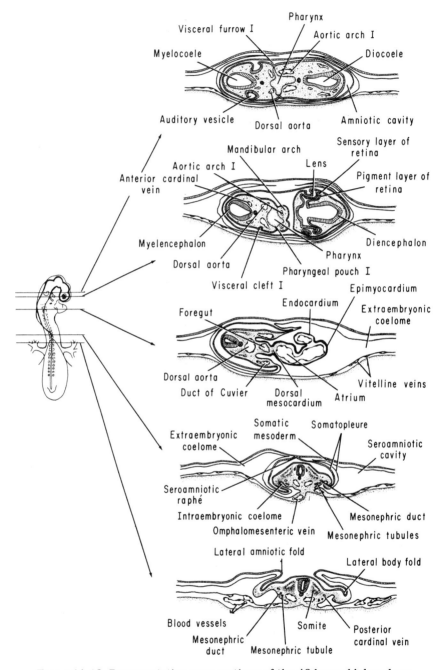

Figure 11-18. Representative cross sections of the 48-hour chick embryo.

Ectodermal Development

Shortly after 33 hours, the brain presents a metameric appearance. These segments, since they appear in the brain, are called neuromeres. Eleven definitely can be seen in the post-33-hour chick. Figure 11-19 represents a diagrammatic view of the neuromeric arrangement in the developing brain. Neuromeres 1, 2, and 3 contribute to the development of the prosencephalon. Neuromeres 4 and 5 develop into the mesencephalon. The remaining neuromeres, 6 through 11, are the rudiments of the rhombencephalon. These three brain regions further differentiate into the five adult brain segments. Neuromeres 1 and 2 become the telencephalon; neuromere 3 develops into the diencephalon. The fourth and fifth neuromeres are forerunners of the mesencephalon. The sixth and seventh are precursors of the metencephalon, while the remaining neuromeres, 8 through 11, develop into the mylencephalon.

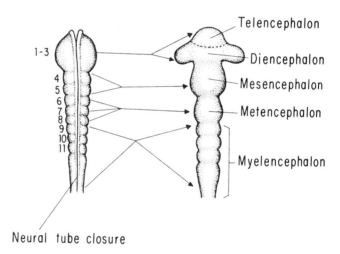

Figure 11-19. Diagrammatic representation of the neuromeric arrangement in the 33-hour chick brain and the adult brain structures into which they develop.

A midsagittal section through a 48-hour chick brain is seen in Figure 11-20. On it a number of important structures can be seen. The most anterior portion of the floor of the brain, that around which the cranial flexure occurs, is the tuberculum posterius. This occurs in the floor of the midbrain region. Anterior to this, a ventral

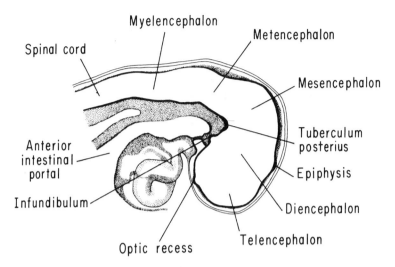

Figure 11-20. Midsagittal section through a 48-hour chick brain showing the more important anatomical structures.

evagination of the diencephalic floor, the infundibulum, occurs. Anterior to this is a depression, the recessus opticus. The optic vesicles are bilateral evaginations of this area. The telencephalon at a later stage will give rise to the left and right cerebral hemispheres. At 48 hours these are not present, and the telencephalon contains only a single cavity, the telecoele. Dorsally, in the region of the diencephalon, a slight evagination occurs. This is the epiphysis, the forerunner of the pineal gland. The midbrain has no distinguishable characteristics at this time. Also, the differentiation between metencephalon and mylencephalon is difficult to see.

The rest of the neural tube develops into the spinal cord. From its method of formation, the inner lining of the neurocoele is nonnervous in character. Surrounding this are the cells of the nervous layer. These are now referred to as neuroblasts. They will differentiate into the functional cells of the nervous system, the neurons, and the supporting elements, the neuroglial cells. The neural crest cords, which have developed from the forming neural tube, begin to segment in a manner similar to that of the somites. These will form the ganglia of the spinal nerves. Consequently, one pair of ganglia is found for each pair of somites. In the brain region, three major cranial ganglia are being formed. The fifth ganglion represents the sensory nerve cell bodies of the trigeminal or fifth cranial nerve. The ganglion of the seventh and eighth cranial nerves occurs just anterior to the developing otic vesicle. This ganglion, the acousticofacialis ganglion, contains the sensory nerve cells of both the seventh and eighth cranial nerves, the facial and the auditory. The ninth and tenth cranial nerves, the glossopharyngeal and vagus, originally have a common ganglion which occurs just posterior to the developing otic vesicle (Fig. 11-17).

At approximately 35 hours, a thickening occurs in the ectoderm at a point opposite the most posterior interneuromeric constriction. These thickenings are the auditory placodes and will eventually invaginate to become the otic vesicles. At the 48-hour stage, this invagination has already occurred, and the presumptive ear can be found as an invaginated vesicle, the otocyst. In the region of the diencephalon the already formed optic vesicles begin to invaginate, giving them the appearance of cups. The innermost layer of the cup is destined to become the retina of the eye. The space between the dorsal and ventral extensions of the cup will be the pupil, and the outer layer of the cup will become the pigmented layer of the eye. At a point opposite the optic cup, the ectoderm of the head invaginates to form the lens. All these structures can be seen in the serial sections represented in Figure 11-18B.

In the ventral region of the head, an ectodermal invagination representing the future mouth area begins to develop. This is the stomadeum. An evagination from the roof of the stomadeum, Rathke's pocket, begins to bud off. This pocket of ectoderm then migrates toward the developing infundibulum of the diencephalic portion of the brain. Together, these two rudiments form the adult pituitary gland or hypophysis. The rudiment formed from the roof of the stomadeum will form the pars distalis, pars tuberalis, and pars intermedia of the adult gland, whereas the infundibular process will form only the pars nervosa.

Gill Arch Development

The visceral pouches and arches of the chick differ from those found in previous vertebrates in that they have no respiratory function. They cannot truly be called respiratory or branchial arches. The terms "visceral pouches" and "arches" will therefore be retained for the study of the chick. The method of formation is similar to that seen in the previous embryos. Each gill slit forms as the result of an evagination of the endoderm of the pharyngeal region of the gut (pharyngeal pouches), and a concomitant invagination of the ectoderm of the head (visceral clefts). The solid areas between successive gill slits are the visceral arches. At the 48-hour stage, four pairs are developed. The first arch, the mandibular arch, is well formed, as are the second or hyoid, the third, and the fourth. Within each of the arches an aortic arch develops.

Endodermal Development

The gut continues to elongate as a result of the continuing caudal progression of the subcephalic fold. The anterior portion of the gut, i.e., the portion anterior to the anterior intestinal portal, is referred to as the foregut. It is in this region that most of the early development occurs. In the floor of the pharynx, in a position midway between the two second pharyngeal pouches, a ventrally directed evagination occurs. This is the forerunner of the thyroid gland. Posterior to the pharyngeal pouches, another ventral evagination, the laryngotracheal groove, appears. Following this posteriorly, it can be seen that the laryngotracheal groove bifurcates. This bifurcation represents the rudiments of the two lungs. The laryngotracheal groove is the forerunner of the larynx and the trachea, whereas the primordial lung buds will form the bronchii, the lungs, and the internal ramifications of the respiratory tubules. The opening of the laryngotracheal groove into the foregut will be called the glottis in later stages. Just anterior to the anterior intestinal portal, another pair of evaginations, directed anteriorly, occurs. These dual evaginations, the anterior and posterior diverticula are the primordia of the liver. These liver diverticula give rise not only to the liver, but also to the gallbladder and to the various ducts associated with these glands. The midgut is still the most extensive portion of the digestive tract. It consists of the entire portion between the anterior and posterior intestinal portals. This region of the gut still has no floor and is open directly to the underlying yolk. Posteriorly, a subcaudal fold begins to lift the tail from the yolk in a manner very similar to that which occurred in the anterior region of the embryo. This subcaudal fold creates both a posterior intestinal portal and a hindgut.

In the anterior region of the foregut an evagination appears, which is directed toward the developing stomadeum. This is the oral outpocketing. This oral outpocketing meets the stomadeal invagination and forms a two-layered structure, the oral plate. When this ruptures, the mouth is formed. The region anterior to the foregut evagination is the preoral gut. A similar ectodermal invagination occurs in the posterior region of the embryo. This is the proctodeum.

An evagination of the hindgut meets this in a fashion similar to that found in the anterior region. This double layer is the anal or cloacal plate. When this ruptures, the anus is formed.

Mesodermal Development

The original mesodermal sheath, already segmented into dorsal epimere, intermediate mesomere, and lateral hypomere, further differentiates by 48 hours. The somites are already separated into an outer dermatome and an inner myotome, separated by a cavity, the myocoele. The inner edges of the myotome begin to bud off cells, sclerotomal cells, which will form the axial skeletal structures. The intermediate mesoderm also continues to develop. This will be seen, in all the later stages, to contribute to the development of the excretory system. At this stage the pronephros develops. The pronephros, or head kidney, is a nonfunctional kidney. Pronephric tubules, which are connected to the coelomic cavity, enter into a pronephric duct. This pronephric duct continues posteriorly until it enters the cloaca. A diagrammatic representation of the early pronephros can be seen in Figure 11-21. The mesonephros, the second kidney to appear chronologically, develops later. The hypomere undergoes rather extensive development, both embryonic and extraembryonic. Embryonically, one of the most conspicuous structures at the 48-hour stage is the heart. It actually began to develop prior to the 33-hour stage. Discussion of it was, however, postponed so that the complete study of the developing heart through the 48-hour stage could be presented.

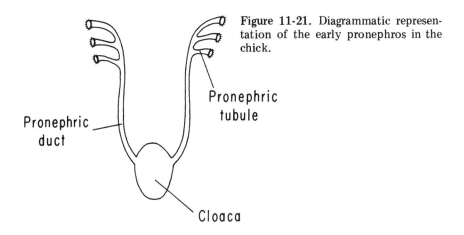

Figure 11-21. Diagrammatic representation of the early pronephros in the chick.

As the splanchnic mesoderm courses down and around the gut region, it appears as in Figures 11-16 and 11-17. This opening within the splanchnic mesoderm, bounded dorsally by the dorsal mesentery and ventrally by the ventral mesentery, is the site of formation of the heart. Cells, budding off from the splanchnic mesoderm on each side, form a pair of hollow tubes, the endocardial tubes. These tubes, arising in a dual fashion, are seen to fuse as their development progresses. The cells of the endocardial tubes will form the endocardial lining of the heart, whereas the remainder of the splanchnic mesoderm surrounding the endocardial tube forms the epimyocardium, destined to give rise to the muscle of the heart and the outer lining, the visceral pericardium. Diagrammatically, this early appearance of the heart, viewed dorsally, appears as in Figure 11-22. When this stage is reached, both the dorsal and ventral mesentery degenerate, leaving a tube swinging rather freely within the pericardial space, the name given for the coelomic cavity in this region.

If we now use an analogy to explain the developing heart, we can appreciate the various curvatures found in the 48-hour chick

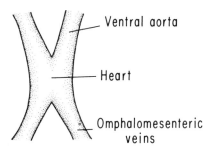

Figure 11-22. Early development of the heart. The two endocardial tubes fuse for a short distance. This area is the heart. The tubes remain separated both anterior and posterior to the heart. Anteriorly they are known as the ventral aortae, while posteriorly they become the omphalomesenteric veins.

Figure 11-23. Heart development analogy. If two points, A and B, are fixed, and a length of string is stretched between them, a mechanical analogy to the developing heart is obtained. These two points represent the anterior and posterior extent of the pericardial cavity, whereas the string represents the endocardial tube. If the string length is to be increased, it must fold upon itself. This creates a figure similar to the developing heart seen in Figure 11-24.

heart. If two points (Fig. 11-23), A and B, are fixed, and a length of string is stretched between them, we have a mechanical analogy to the developing heart at the 33-hour stage. These two points represent the anterior and posterior extent of the pericardial cavity, whereas the string represents the endocardial tube. The only way to increase the length of the string without altering the position of the points A and B is to fold the string upon itself. This is exactly what happens in the development of the heart. As the heart, or endocardial tube, lengthens, it swings to the left and doubles upon itself. Each area of the endocardial tube then begins to differentiate until the structure seen in Figure 11-24 appears. Due to this folding, the atrium which was posterior now appears anterior, and the ventricle which was anterior now appears in a more posterior position. The heart, at this

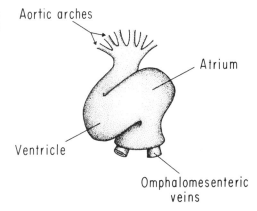

Figure 11-24. Appearance of the heart after the endocardial folding has occurred.

stage, is considered to be a two-chambered heart. It has one atrium and one ventricle. Immediately posterior to the atrium is another large chamber, the sinus venosus. It is usually not counted when vertebrate hearts are discussed as being two-chambered or three-chambered. It does, however, play an important role in the circulatory system. It receives the blood from the left and right common cardinals which drain both the anterior and posterior regions of the body. The endocardial tubes remain fused for a short distance posterior to the sinus venosus as the ductus venosus (Fig. 11-24). The posterior portion of the developing endocardial tube remains double. This double portion constitutes the omphalomesenteric veins, which go out to the surface of the yolk. Anteriorly, the endocardial tube remains double as the ventral aortae. These course beneath the pharyngeal area. The aortic arches, which have developed within the visceral arches, connect these ventral aortae to the dorsal aortae on each side of the body, completing the aortic arch structure. At the 48-hour stage, three aortic arches develop.

Other Circulatory Vessels

The loosely arranged mesodermal cells found throughout the body as mesenchymal cells begin to organize themselves into distinct blood vessels. The posterior extension of the double anterior dorsal aortae is the single dorsal aorta, which courses posteriorly throughout the rest of the embryonic body. In the region of the midgut, a pair of large segmental arteries appears. These are the omphalomesenteric arteries, which course out to the surface of the yolk. This completes the extraembryonic yolk circulatory pattern. A drop of blood, starting in the area of the yolk, would pass through the circulatory structures in the following order:

1. yolk capillaries
2. vitelline veins
3. omphalomesenteric veins
4. ductus venosus
5. sinus venosus
6. atrium
7. ventricle
8. conus arteriosus
9. bulbous arteriosus
10. ventral aorta
11. aortic arches
12. dorsal aorta
13. omphalomesenteric arteries
14. vitelline arteries
15. yolk capillaries

On each side of the developing embryonic body other vessels begin to appear. These are the cardinal veins, which have already been seen in other vertebrates. The vessels that appear anterior to the developing heart are the left and right anterior cardinal veins; those that develop posterior to the developing heart are the left and right posterior cardinals. They meet in a region just dorsal to the sinus venosus and form the single common cardinal vein on each side of the body. These left and right common cardinal veins (also called the ducts of Cuvier) enter the sinus venosus. With the establishment of the cardinal system, a complete embryonic circulatory pattern is established.

The surface of the yolk, at this stage, also becomes patterned with a series of blood vessels. These can be seen in Figure 11-25. Two primary sets of veins, which connect to the omphalomesenteric veins, develop. These are the anterior, posterior, and lateral vitelline veins. These become more extensive as the embryo develops. Surrounding

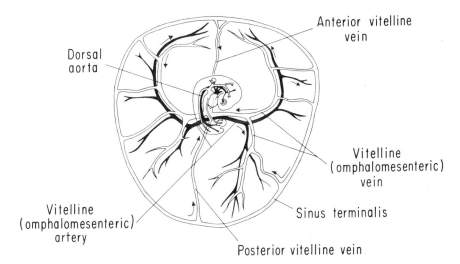

Figure 11-25. Yolk circulatory pattern. This complex of veins and arteries develops over the surface of the yolk. Nutritive material is carried to the embryo through these channels.

the area vasculosa is the sinus terminalis which connects to these vitelline veins. This elaborate extraembryonic circulatory system is necessary when one considers that the yolk is the sole source of nutrient supply for the developing embryo for the entire period of 21 days.

The Cleidoic Egg

At this point, it is necessary to explain the development of the extraembryonic membranes that distinguish the eggs of the reptiles, birds, and mammals from those of other vertebrates. These eggs are so characteristic that the popular term "amniote" is applied to vertebrates of these three classes. The five lower classes of vertebrates, not possessing the extraembryonic membranes, are frequently referred to as anamniotes. It was mentioned in Chapter 10 that the cleidoic egg would be studied in the present chapter. The following discussion of the extraembryonic membranes, therefore, applies equally as well for the reptiles as for the mammals discussed in Chapter 12.

There are basically four extraembryonic membranes that develop during the early phases of embryology. These are the amnion, chorion, allantois, and yolk sac. The extraembryonic extensions of the somatopleure and splanchnopleure are the tissues that take part in the formation of these membranes. These membranes begin to form early during the second day of development and continue to develop until approximately the fifth day. The amniotic fold is the first to appear. This actually is four major folds. One occurs in the head region, one in the tail region, and a pair occurs on each side of the body. These are referred to as the cephalic, caudal, and lateral body amniotic folds, respectively. The inner lining of the newly formed amniotic cavity is the amnion. It is composed of the ectoderm and somatic mesoderm. As the amniotic folds fuse, the resultant degeneration creates also an outer layer, the chorion. It is also, therefore, composed of ectoderm and somatic mesoderm. This chorion layer continues around the egg, just inside the shell membranes. The cavity between the embryo and the amnion is the amniotic cavity and is filled with amniotic fluid, a fluid that is isotonic with the body of the chick. Many embryologists have referred to this as the embryo's "private swimming pool." The cavity between the amnion and chorion is the chorionic cavity.

The splanchnopleure continues down and around the yolk substance. This forms the yolk sac. The ventralmost extension of the yolk sac, where the two edges of the yolk sac have not yet approached one another, is referred to as the yolk sac umbilicus. Soon, however, this also is obliterated and each side of the yolk sac fuses with its mate from the other side. A cross section through an embryo at this stage is seen after the three membranes, the amnion, chorion, and yolk sac, have developed (Fig. 11-26).

A fourth cavity, the allantois, develops from an evagination of the hindgut region. This evagination of the hindgut pushes the splanchnopleure out in front of it. This gradually increases in size, eventually filling all the chorionic cavity. This allantois serves as an embryonic lung and chamber for metabolic waste products. The completed embryo, with all its extraembryonic membranes, is represented diagrammatically in Figure 11-26. From this diagram it can be seen that there is a very close relationship between the allantois and the chorion. This relationship is so intimate that the dual membrane is often referred to as the chorioallantoic membrane. Since it is very heavily vascularized, it is frequently used by experimental embryologists as a site for transplantation of tissues. The blood vessels formed in the splanchnopleure are now found next to the shell. In this position they are ideally located for the diffusion of oxygen and carbon dioxide. In this way the allantois functions as an embryonic lung.

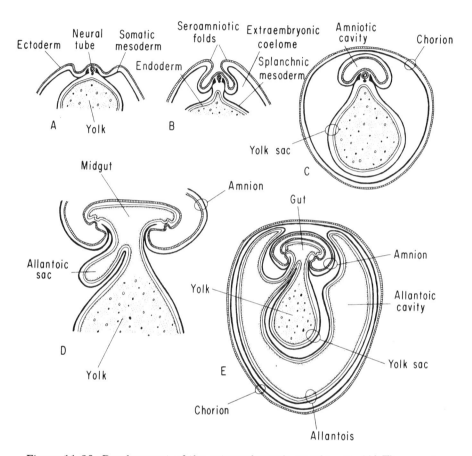

Figure 11-26. Development of the extraembryonic membranes. (A) The somatopleure folds upward, above the embryo, as the seroamniotic folds. (B and C) These seroamniotic folds eventually meet above the embryo, fuse, and break apart, forming an inner amnion and an outer chorion. The splanchnopleure continues to grow down and around the yolk as the yolk sac. (D) An evagination of the hindgut, the allantois, grows out and becomes covered with splanchnic mesoderm. Eventually this allantois fills the entire chorionic cavity. The embryo after the formation of these four extraembryonic membranes is seen in D.

The 72-Hour Chick

Flexures and Torsion

Figures 11-27 and 11-28 show both the whole mount and selected serial sections of the 72-hour chick. From the whole mount it can be seen that the flexion, which started between the thirty-third and forty-eighth hour of incubation, has now continued to the point that the embryo looks like a figure C. Four flexures are present. The first is the original cranial flexure. This occurs in the region of the mesencephalon. The second, in the region of the neck, is the cervical flexure. The third is a reverse type of flexure found in the lumbar region. It is designated the lumbar flexure. The last one is the caudal flexure in the region of the tail.

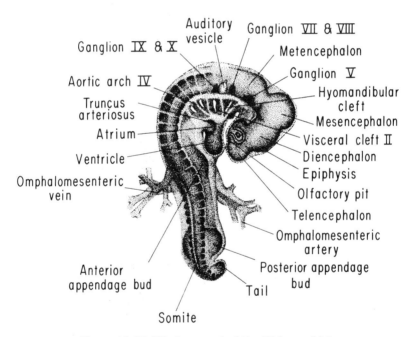

Figure 11-27. Whole mount of the 72-hour chick.

The other process, torsion, which began shortly after 33 hours of incubation, has continued so nearly the entire embryo now lies on its left side on the yolk. The somites, which were so bilaterally apparent in the earlier stages, particularly in the posterior region, now appear with only the right somites visible from a dorsal view.

Figure 11-28. Selected cross sections of the 72-hour chick embryo.

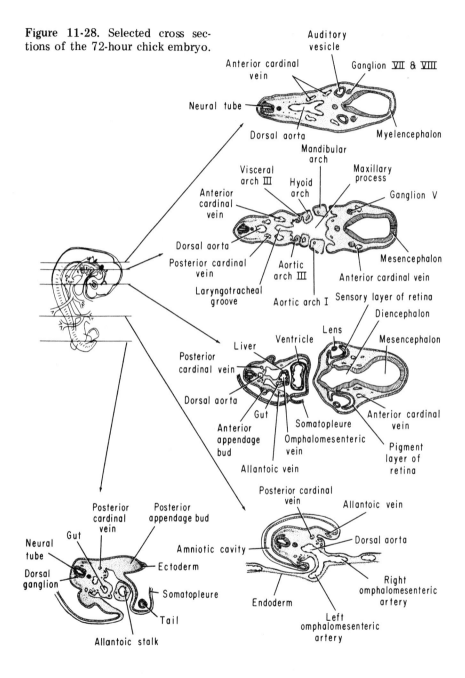

As a result of the flexures, the hindbrain is now the most anterior portion of the embryo. It is difficult, when studying serial sections, to realize that one is progressing anteriorly in the head region since the slides progress posteriorly.

Another conspicuous pair of structures is now apparent in the 72-hour chick. These are the anlages of both the anterior and posterior appendages. They are referred to as the anterior limb bud and posterior limb bud. The anterior limb buds will eventually develop into the wings, whereas the posterior limb buds will develop into the legs of the chicken. Figure 11-28E shows a section through one of these limb buds. As can be seen, the epidermis in this area is thickened over the distal end of the bud. This thickening is frequently called Saunders' ridge. The interior of the limb bud is still composed of loosely organized mesenchymal cells. These mesenchymal cells will, in the future, develop into the bones, muscles, blood vessels, and connective tissue of the adult limbs.

The Nervous System

The Brain

The most anterior portion of the brain, the telencephalon, up to this time, has been a single cavity. By the end of the third day of incubation, the original telencoele has started to divide into two separate cavities. These cavities represent the first and second brain ventricles. They are connected to the single cavity of the diencephalon by an opening, the foramen of Munro. The two portions of the telencephalon that are now apparent are the forerunners of the cerebral hemispheres.

The diencephalon has also undergone some changes since the 48-hour stage. The roof has put forth an evagination, the epiphysis or pineal gland. The walls of the diencephalon appear thicker, representing the primordia of the thalamic portion of the brain. Ventrally, the infundibulum which has already been seen, continues to lie in close proximity to Rathke's pocket. These will form, together, the pituitary gland of the adult. The cavity of the diencephalon, the diencoele, is the third brain ventricle.

The mesencephalon develops in a manner somewhat different from the rest of the brain. Instead of the cavity enlarging, it is actually narrowed. This narrowing is due to an increase in the thickness of both the wall and floor of the mesencephalic region. These thickenings, the crura cerebri, will contain nerve fibers that are ascending and descending along the brain pathways. The narrow cavity, now termed the cerebral aqueduct, or the aqueduct of Sylvius, will serve to connect the third brain ventricle in the diencephalon with the fourth brain ventricle in the myelencephalon. The roof of the mesencephalon is the most conspicuous in terms of its development. It forms a four-lobed body, the corpora quadrigemina, composed of an anterior pair of projections, the superior colliculi, and a posterior pair of projections, the inferior colliculi. These represent reflex centers of vision and hearing, respectively.

The metencephalon is rather poorly developed at this stage, although the roof is somewhat thicker than in the more posterior myelencephalon. This thickened roof is the forerunner of the cerebellar hemisphere. The floor, which is also somewhat thicker, is the primordium of the pons varolii. The cavity within the metencephalon remains rather narrow. It will become an extension of the cerebral aqueduct.

The myelencephalon, the most posterior portion of the brain, contains within it an extremely large cavity, the myelencoele. This myelencoele is, in reality, the fourth brain ventricle. It is located between the cerebral aqueduct anteriorly and the neurocoele of the neural tube posteriorly. The roof of the myelencephalon is extremely thin. This will eventually develop into the posterior choroid plexus.

The Spinal Cord

The neuroblasts in the spinal cord continue to proliferate, forming condensations that are heavier in the dorsal and ventral areas of the neural tube than they are in the lateral region. This produces the dorsal and ventral horns of the gray matter. A few axonic fibers of the association neurons are found in the area immediately surrounding the original neural tube. These form the white matter of the spinal cord. Development does not proceed much beyond this at the 72-hour stage.

Peripheral Nervous System

The peripheral nervous system takes the form of embryonic cranial and spinal nerves. It is helpful at this point to recall that the nerve cell bodies of the motor neurons are found within the central nervous system, that is, within the brain or spinal cord, whereas the nerve cell bodies of the sensory neurons are found outside the central nervous system. In this embryo they will be found in dorsal spinal ganglia in the spinal cord, and in cranial ganglia in the brain.

The cranial ganglia present at the 72-hour stage are the trigeminal or ganglion of the fifth cranial nerve, which is found lateral to the mesencephalon. The ganglion of the seventh and eighth cranial nerves is, at this stage, still nonsegmented and is found immediately anterior to the auditory vesicle. The ganglion of the ninth and tenth cranial nerves is found also nonsegmented just posterior to the auditory vesicle.

The ganglia of the spinal nerves are found dorsal-lateral to the spinal cord. These are arranged in a segmented fashion, much the same as the myotomes were. One pair of spinal nerves will emerge, eventually, between each consecutive pair of vertebrae. They are seen in cross section as distinct condensations of neural crest cells.

Sense Organs

The Otic Vesicle

The otic vesicle develops as an invagination of the superficial ectoderm in a region lateral to the myelencephalon. This invagination buds off to form two complete vesicles lying in this region. No further development of the otic vesicles takes place up to 72 hours.

The Nasal Pits

The nasal placodes, which were present at the 48-hour stage as thickenings in the ectoderm in the region of the forebrain, have now invaginated to form definite olfactory pits. The nervous ectoderm of the olfactory pits soon will differentiate into the sensory cells of the olfactory nerve.

The Eye

The invaginated optic cup, lying beneath the overlying ectoderm, causes a change in that ectoderm. An invagination which soon buds off, the lens, is formed. This formation of the lens is one of the classic examples of induction. The underlying developing optic cup in some way affects the development of the overlying ectoderm, for if the optic cup is removed, the lens fails to develop. In amphibians, if the optic cup is transplanted to an ectopic site, such as beneath the ectoderm of the ventral abdominal wall, a lens will form in this ectoderm. This type of influence of one tissue on another is referred to as induction.

Endodermal Development

A number of changes occur in the digestive tube during the period from the forty-eighth to seventy-second hour of incubation. These include the breaking through of the oral plate, changes in the structure of the aortic arches, thyroid gland evagination, development of the primordial respiratory system, differentiation of the esophagus and stomach, formation of the ductus choledochus, further differentiation of the gallbladder and liver, and the establishment of a distinct hindgut region. Each of these will be discussed separately.

The Oral Plate

It has already been seen that in the anterior region of the foregut an evagination occurs which meets the stomadeal invagination of the ectoderm. The double-layered membrane that forms is the oral plate. This oral plate has remained intact up to and through the 48-hour stage, but once the 72-hour stage is reached, the oral plate ruptures, forming a direct opening from the exterior into the gut. This opening is the mouth. It should be noted, therefore, that the mouth is of a dual origin. The more external portion of the mouth is ectodermal, coming from the stomadeal invagination, whereas the internal portion of the mouth is endodermal, coming from the endodermal outpocketing of the anterior portion of the foregut. The evagination of the foregut does not occur at the extreme anterior end of the gut but at a short distance posterior to it. This leaves a blind pocket proceeding anteriorly from the mouth area. This is a transitory structure, but it is known in the embryo as the preoral gut or Seessel's pocket.

Pharyngeal Changes

The original visceral arch structure has not changed much since the 48-hour chick. The first cleft does break through, however, although this gill slit is never functional. The thyroid gland, which has already made its appearance in the floor of the pharynx between the two second pharyngeal pouches, continues to deepen and forms a definite evagination by this stage. Posterior to the last of the pharyngeal pouches, the pharynx deepens somewhat to form a groove, the laryngotracheal groove. This laryngotracheal groove is the forerunner of the larynx and trachea. If it is followed posteriorly, the laryngotracheal groove is seen to separate from the pharyngeal area and bifurcate posteriorly into two lung buds. The coelomic cavity surrounding these lung buds is the pleural cavity. They can be seen in Figure 11-29. Posterior to the pharyngeal region, the foregut narrows considerably. In the pharyngeal area it was compressed dorsoventrally and expanded laterally. This was due to the formation of pharyngeal pouches. In the region immediately posterior to the pharyngeal area, this narrowing is very conspicuous. In this area, the foregut is now called the esophagus. If this esophagus is followed posteriorly, it will soon be seen to enlarge. At this enlargement, the foregut is called the stomach. This original single enlargement will eventually become the proventriculus and gizzard of the adult bird.

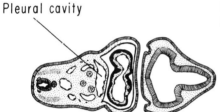

Figure 11-29. Cross section of a chick embryo showing the pleural cavity.

Just anterior to the anterior intestinal portal, the liver diverticula can be seen to leave from a common evagination of the ventral portion of the gut. This common evagination is the ductus choledochus. The anterior liver diverticulum proceeds on the left side of the ductus venosus, whereas the more posteriorly located right primordium is found to course on the right side of the ductus venosus. These do not remain separate, however, and anastomoses are found rather frequently to occur between the left and right liver primordia. At their most anterior end, the gallbladder develops from the posterior lobe as an evagination. The connection between this gallbladder evagination and the posterior lobe is a thin duct, the cystic duct. The original primordia of the liver will give rise to the common bile ducts, hepatic ducts, cystic ducts, liver lobes, and gallbladder.

The hindgut is formed in a manner very similar to that of the foregut. An undercutting, the subcaudal fold, lifts the tail bud from the blastoderm. The posterior portion of the gut, therefore, attains a floor. The opening into this posterior region of the gut is the posterior intestinal portal. At the 72-hour stage, the hindgut is very small

The 96-Hour Chick

During the first 72 hours of development, most of the organ anlage are already established. The period of development from five days to hatching is now primarily one of organogenesis or further differentiation of these organs. This more extensive development does result, however, in an embryo with an altered appearance. Figure 11-30 represents a whole mount of a 96-hour embryo.

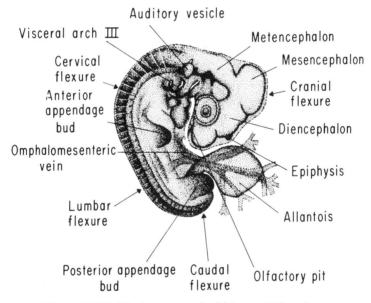

Figure 11-30. Whole mount of a 96-hour chick embryo.

It will be noted that the flexures and torsion are now much more complete than they were at the 72-hour stage. Torsion, which began at approximately 48 hours, has now reached a point at which the entire embryo is lying with its left side toward the yolk. The more extensive flexion in the cephalic, cervical, lumbar, and caudal regions gives to this embryo a C-shaped appearance. The limb buds, which were just making their appearance at 72 hours, are now fairly extensive structures and can be readily seen.

Somite Differentiation

The somites, which already had differentiated into dermatomal, mytomal, and sclerotomal regions, now continue their internal differentiation. The cells of the myotomes begin to take on a rather elongated position, indicative of their future development into striated muscle. The dermatome lies at this stage as a band of tissue just beneath the epidermis. This dermatome ultimately gives rise to the dermis of the adult chick. The sclerotomal cells have left their site of formation and have started their migration around the notochord and neural tube. Here they will form the basic arcualia of the adult vertebrae. The development of the vertebrae in the birds and mammals is very similar to that already described under amphibian development.

Foregut Derivatives

Only those areas of the foregut that have not been treated extensively for the 72-hour chick will be discussed here. The thyroid gland, which was seen at the last stage to be simply an evagination in the floor of the pharynx between the second pharyngeal pouches, continues to push ventrally and becomes a bilobed structure. In the adult, the thyroid and pituitary glands are intimately associated physiologically. Thyrotropic hormone from the pituitary gland activates or stimulates the thyroid gland to produce its hormone, thyroxin. This has led investigators to wonder whether the pituitary gland influences the development of the embryonic thyroid. By removing the pituitary gland at an early stage, it has been shown that the thyroid gland will stop developing at 12 days. In other words, the early development of the thyroid is not under the influence of the pituitary, but the later functional stage of thyroid gland development is under the control of the pituitary hormones.

The respiratory tract, which at the 72-hour stage was called the laryngotracheal groove, has now further differentiated into a series of structures. The opening from the pharynx into the respiratory passageway is the glottis. This is followed immediately by the larynx. In the mammals, the vocal cords develop in this region. The larynx is followed more posteriorly by the trachea. At the 96-hour stage, no substantial histological development of the trachea has taken place. In later development, mesenchymal cells surrounding the trachea differentiate into chondroblasts and, ultimately, chondrocytes. These cartilage cells form segmented rings of hyaline cartilage around the trachea. Posterior to the trachea are bifurcations representing the lung buds. This development of the respiratory structures is occurring at the same time that the aortic arch plan is being modified. The derivative of the sixth arch, the pulmonary artery, soon becomes intimately associated with the developing lung buds.

As one proceeds posteriorly from the region of the pharynx, the digestive tube narrows. This narrow portion is the esophagus. Following this posteriorly it will once again be seen to widen considerably. This pouchlike widening of the foregut is the stomach area. Between the stomach and the anterior intestinal portal, the gut is referred to as the duodenum. It is in this duodenal portion that the common bile duct, the common duct for the pancreas, liver, and gallbladder ducts, enters. The liver diverticulum continues to develop. The right lobe of the liver diverticulum is larger than the left. The pancreatic diverticulum is represented at this stage by three separate evaginations of the duodenal region. One of these is dorsal, opposite the posterior liver diverticulum; two of them are lateral, stemming directly from the ductus choledochus. Immediately above the dorsal pancreatic diverticulum, cells can be seen budding off from the dorsal mesentery. These mesentery cells will differentiate into the adult spleen.

The midgut, which is still open to the yolk, is the future small intestine region of the adult chick. Posterior to the midgut is the hindgut. This, in part, develops into the cloaca. At the 96-hour stage this cloacal region can be seen to receive not only the intestine but also the wolffian or mesonephric ducts of the excretory system.

Development of the Heart

In the 72-hour stage, the heart was seen to be composed of a sinus venosus, atrium, ventricle, and conus arteriosus. The atrial and ventricular portions of the heart now undergo more extensive development. This varies with the vertebrate. In the reptile, the heart is three-chambered, that is, it has two atria and a ventricle which is partly divided. In the birds and mammals, the heart is four-chambered, composed of a left and right atrium and a left and right ventricle.

In order to reach this advanced stage of development, the original heart cavities enlarge. The canal between the atrium and ventricle is referred to as the atrioventricular canal. A partition develops between the left and right side of the heart. That portion of the partition which falls between the left and right side of the original atrial cavity is called the interatrial septum, whereas that found in the original ventricle is the interventricular septum. The ventral end of the interatrial septum is more extensively developed as a cushion septum. The original atrioventricular canal now becomes two canals, a right atrioventricular canal and a left atrioventricular canal. The right canal is soon guarded by the developing tricuspid valve, whereas the left canal is guarded by the bicuspid or mitral valve. This heart structure can be seen in Figure 11-31.

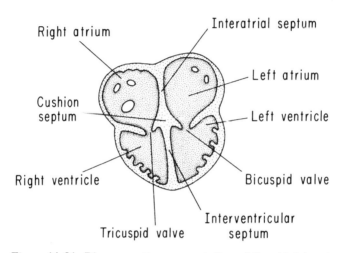

Figure 11-31. Diagrammatic representation of the chick heart.

The Fate of the Aortic Arches

The original aortic arches and their fate have already been described under amphibian development. However, a change does occur in the reptiles, birds, and mammals which must be discussed here. Figure 11-32 represents the fate of the aortic arches in all three vertebrate forms. In all three forms, the first and second aortic arches degenerate. The ventral aorta becomes primarily the external carotid artery, whereas the dorsal aorta and third aortic arch become the internal carotid. The dorsal aorta between the third and fourth arch degenerates, whereas the ventral aorta between the third and fourth becomes the common carotid artery. The fourth arch remains intact in the reptiles, resulting in both a left and right systemic aortic arch. In the birds, the left fourth aortic arch and that portion of the dorsal aorta associated with it degenerate, leaving only the right systemic arch. In the mammals, the right fourth aortic arch and the dorsal aorta associated with it degenerate, leaving only the left systemic arch. The fifth arch degenerates in all three forms, and the sixth arch gives rise to the pulmonary artery. That portion of the sixth arch between the origin of the pulmonary artery and the dorsal aorta is known as the ductus arteriosus, or the duct of Bottalus. This degenerates at birth and acts, in the embryo, as a shunt for the blood, since the lungs are not functioning during the period of embryonic development.

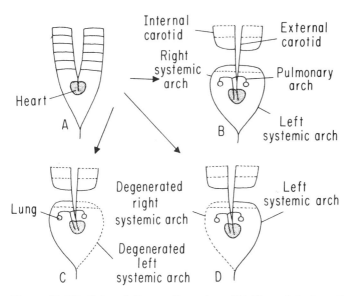

Figure 11-32. Fate of the aortic arches. (A) The original aortic arch structure. (B) The reptilian pattern. (C) The final structure in the birds. (D) The mammalian aortic arch derivatives.

Blood Vessel Development

The dorsal aorta, as it courses through the body, gives off segmental arteries at each somite level. The eighteenth segmental artery enlarges to become the primary subclavian artery. This goes to the developing anterior limb buds. However, as the aortic arch structure is modified, the primary subclavian develops another connection to the third aortic arch. This new vessel becomes the permanent subclavian, and the original eighteenth segmental artery remains simply to nourish the muscles in that region of the body. This development of the subclavian artery can be seen in Figure 11-33.

Figure 11-33. Development of the subclavian artery. (A) The eighteenth segmental artery enlarges to become the primary subclavian artery. The bud of the permanent subclavian artery can be seen leaving the third aortic arch. (B) Completed permanent subclavian artery.

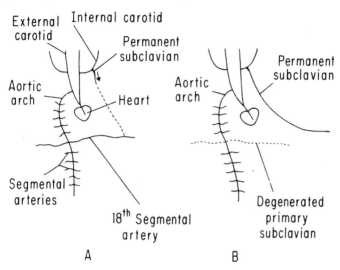

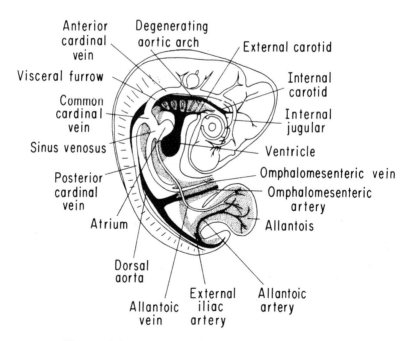

Figure 11-34. Circulatory pattern of the 96-hour chick embryo.

The dorsal aorta also gives off a pair of large arteries to the developing hindlimb buds. These are the sciatic arteries. Anterior to the sciatic arteries are the allantoic or umbilical arteries which go out to the yolk sac and allantois. Although they develop with bilateral symmetry, the right one degenerates, leaving only a single umbilical artery.

The veins develop in a manner very similar to that already seen in the amphibian. The anterior cardinal veins and the duct of Cuvier now become the jugular veins. A new vein in the floor of the mouth develops into the external jugular, whereas the original anterior cardinal becomes the internal jugular. The posterior vena cava develops in a manner very similar to that already seen in the frog. The complete circulatory pattern, shortly after 96 hours, can be seen in Figure 11-34.

Five Days to Term

As a result of the method of embryo formation in the chick, the head, or anterior region of the body, has had more chance to develop than the more posterior regions. Consequently, more than one-half of the four-day chick is head. This necessitates a different rate of growth during the last 17 days of development. The body grows much more rapidly than the head. This phenomenon of differential growth is referred to as heterogonic growth.

The limb buds, which arose at 72 hours, continue to grow. As they do, the internal development becomes more apparent. The mesenchymal cells lying in the central axis of the limbs begin to differentiate into precartilage or chondroblasts. These chondroblasts become organized into definite skeletal structures, and the limbs become jointed. The bones in the limb of the chick are very similar to those found in the reptile and mammal. From a proximal-distal direction, these are the humerus, radius and ulna, carpals, metacarpals, and phalanges. The complete limb can be seen in Figure 11-35.

The yolk sac, which was so dominant in the early stages of development, gradually is absorbed as the developing embryo utilizes this yolk for its metabolic needs. As this yolk sac is absorbed, the tissue forming the yolk sac becomes part of the digestive tract. At birth, the yolk sac is completely absorbed, and the ectoderm which originally surrounded it becomes the ventral body wall. This yolk sac absorption is shown diagrammatically in stages in Figure 11-36.

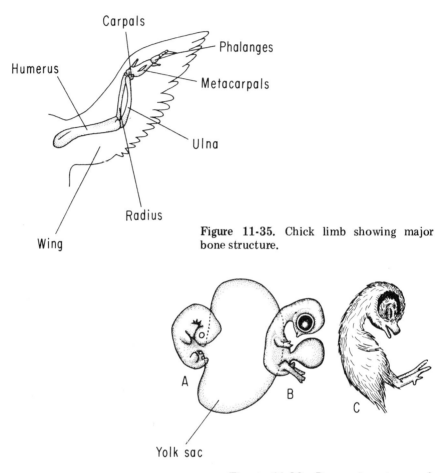

Figure 11-35. Chick limb showing major bone structure.

Figure 11-36. Progressive stages of yolk sac resorption.

The body surface of the chick differs from that of the reptile and mammal in that the avian body is covered with feathers. The reptilian body is covered with scales, and the mammalian body is covered with hair. In the formation of feathers the epidermis invaginates and, secondarily, evaginates to form a feather germ. A condensation of blood vessels beneath this developing feather germ grows into the developing feather as the feather pulp. The position of the feathers is also genetically controlled and follows a well-established pattern. The development of the feather as well as a small section of skin in which feathers are developing is shown in Figure 11-37.

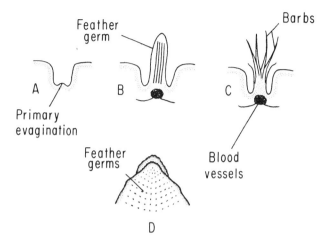

Figure 11-37. Development of the feather germ. (A) The epidermis invaginates and secondarily evaginates to form a feather germ. (B) This becomes underlain by a condensation of blood vessels (a glomerulus). The basic feather differentiation begins. (C) Further differentiation of the feather. (D) A section of the skin of the dorsal region of the chick showing the patterned arrangment of the feather germs. Each dot is a feather germ.

It might be mentioned at this time that bone develops in two different ways. In one, the mesenchymal cells ossify directly; that is, they become osteoblasts and osteocytes. This type of bone formation results in dermal bone. Very few bones of the avian body are of this type. A few in the pectoral girdle and some of the covering bones of the skull are dermal bones. The same is true of reptiles and mammals. The second type of bone is cartilaginous bone. In this type of development, the bone is first formed as cartilage. The mesenchymal cells develop into chondroblasts which further differentiate into chondrocytes. This cartilaginous bone later ossifies as true bone. Most of the bones of the reptilian, mammalian, and avian body are of this type.

The skull forms first from mesenchymal cells which condense around the developing brain. As these form a cartilaginous structure, this original brain case is referred to as the chondrocranium. Later in development this chondrocranium ossifies and is then referred to as the neurocranium. This is an example of cartilaginous bone. Other cells from the dermis organize on top of the neurocranium and form flattened head bones. These flattened head bones form the dermatocranium of the skull. When a vertebrate skull is examined, it is usually the dermal bones that are seen. The bones of the gill arches are called the splanchnocranium. In the reptiles, birds, and mammals the splanchnocranium or, as it is sometimes called, the visceral skeleton, later degenerates, except for a few bones such as the hyoid bone, ear ossicles, and laryngeal cartilages.

The Respiratory Tract

The original laryngotracheal groove continues to undergo further development. The opening from the pharynx into the respiratory tract is the glottis. This leads directly into the larynx, the most anterior portion of the respiratory tract. Posterior to this is the trachea, characterized by cartilaginous rings. These cartilaginous rings represent mesenchymal depositions around the original endodermal lining. Posteriorly, the trachea bifurcates into two bronchii which lead into the lungs themselves.

The lung structure of the chick, as well as of all birds, is different from either the reptile or mammal. The high metabolic requirements of the bird require an almost constant source of oxygen. In the typical breathing pattern, as established by both reptiles and mammals, a period of exhalation reduces the efficiency of the lungs by nearly one-half. This is overcome in the chick by the development of accessory lungs called air sacs. Five air sacs develop from each primary lung. These are the cervical, clavicular, thoracic, anterior abdominal, and posterior abdominal air sacs. These act in the adult as reservoirs of air. They are endodermal in origin. Concomitant with the development of the lungs is the further development of the pulmonary artery as an outgrowth of the sixth aortic arch. It must be pointed out that respiration is not a function of the embryonic lung. The exchange of gases between the environment and the developing chick takes place through the shell and the chorioallantoic membrane.

The Digestive Tract

The original endodermal lining of the digestive tract now becomes invested with mesenchymal cells. These form a series of consecutive layers around the original endodermal lining. The original lining forms solely the mucosa of the adult digestive tract. The mesenchymal cells form an overlying submucosa. This is followed by a circular and longitudinal layer of muscles. The outside of the digestive tract is lined with the original splanchnic mesoderm, which is now called the visceral peritoneum. A cross section through the gut region is seen in Figure 11-38.

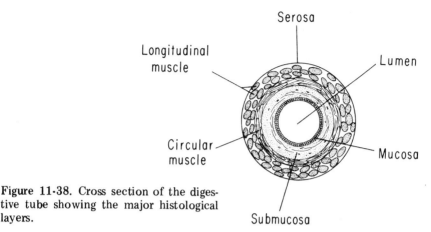

Figure 11-38. Cross section of the digestive tube showing the major histological layers.

Specific modifications of the digestive tract occur in the birds. It must be remembered that the bird has no teeth with which to chew its food, nor does it have a salivary gland to moisten the food before swallowing. These functions, chewing and moistening, must be accomplished by other structures of the digestive tract. An evagination in the esophageal region, the crop, secretes a substance called pigeon milk. This moistens the food as it passes down the esophagus into the stomach. The original single enlargement, which in previous stages was called simply the stomach, now becomes subdivided into an anterior proventriculus and a posterior gizzard. The internal lining of the gizzard becomes cornified in the form of gizzard plates. Food, together with nonnutritive pebbles or stones, falling against the gizzard plates, is reduced to a chyme. The gizzard itself opens directly into the duodenal portion of the gut.

The liver and pancreatic ducts persist, although the ductus choledochus disappears. All three ducts, that is two pancreatic ducts and a liver duct, empty directly into the duodenum itself. Posteriorly, the intestine becomes further modified by the investment of the same layers already seen in Figure 11-38. The posterior termination of the gut is in the region of the cloaca. This opens to the environment through the cloacal opening.

The Circulatory System

The formation of the pre- and post-caval systems, as well as of the hepatic portal venous system, is similar to those already described. The modification of the aortic arches has also been described. Most of the small blood vessels are formed by a coalescence of mesenchymal cells in the area of manufacture. These coalescing tubes soon unite with one another to form definitive blood vessels. The basic circulatory plan of a newly hatched chick can be seen in Figure 11-39.

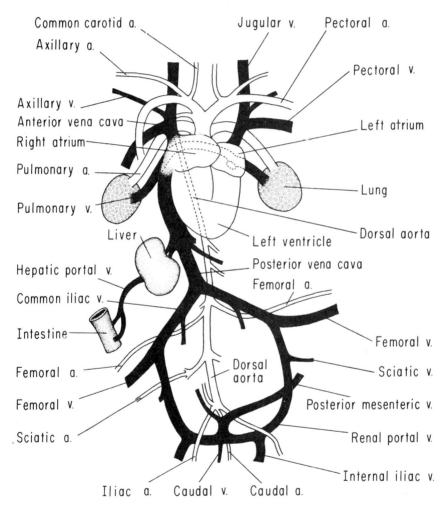

Figure 11-39. Adult avian circulatory pattern.

The Excretory System

The mesonephric kidney, which acts as the functional excretory unit of the developing chick, will be replaced in the adult by the metanephric kidney. This kidney makes its appearance at the 72-hour stage and continue to develop throughout the period of incubation. The metanephric duct is an outgrowth or evagination of the posterior end of the mesonephric duct. As it grows anteriorly, its distal end expands into what will become the pelvis of the kidney (the renal pelvis). The metanephric duct itself will become the ureter. Due, undoubtedly, to inductive influences of the developing renal pelvis, mesenchymal cells in the area begin to coalesce around the pelvic region. These form the renal cortex with all of its histological structure. The details of the development of the cortical and medullary portions of the kidney are beyond the scope of an elementary textbook. The metanephric kidney development is pictured in Figure 11-40.

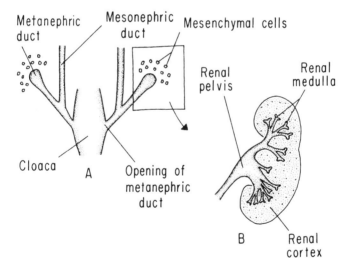

Figure 11-40. Development of the metanephric kidney. (A) The metanephric duct develops as an evagination of the posterior portion of the mesonephric duct. The distal end of the metanephric duct (in square) enlarges as the renal pelvis and induces the surrounding mesenchymal cells to form the kidney cortex. (B) Completed early kidney structure.

The Reproductive System

The reproductive system is so closely associated with the excretory system that they are frequently referred to as a single urogenital system. Much of this relationship goes back to the embryonic stages for, from its beginning, the gonad is intimately associated with the kidney structure. On the ventral and medial surfaces of the mesonephric kidney, the gonads can be seen to arise as a thickening, termed the gonadal ridge. This gonadal ridge develops into the gonads, i.e., either the ovary or testis. The individual cells that are later destined to become the eggs and sperm of the adult female and male, respectively, do not originate in the same way. Actually, there is no universal agreement regarding the origin of the germ cells. It appears fairly certain, however, that they are formed in other parts of the body and migrate to the developing gonads. The ducts carrying the sperm away from the testis are modifications of the original mesonephric duct. The mesonephric duct is no longer utilized by the chick as the metanephros develops, except as a duct for the passage of sperm. These ducts are called, in the adult, the ductus deferens and the ejaculatory duct. They can be seen in Figure 11-41. In the female, two new ducts develop, which connect with

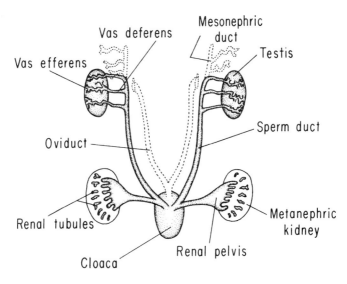

Figure 11-41. Development of the male reproductive system. The posterior portion of the mesonephric duct in the male is the sperm duct or vas deferens.

each developing ovary. These new ducts, the müllerian ducts, fuse with one another posteriorly, forming the uterus. A complete drawing of the female reproductive system is seen in Figure 11-42.

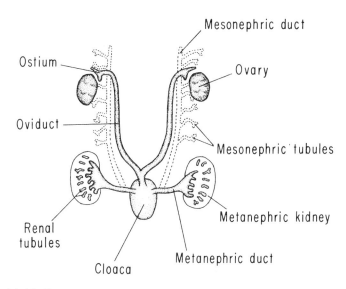

Figure 11-42. Development of the female reproductive system. Two new ducts develop, which connect with each developing ovary. These new ducts, the müllerian ducts, fuse with one another posteriorly, forming the uterus.

"Lo, children are an heritage of the Lord: and the fruit of the womb is his reward. As arrows are in the hand of a mighty man; so are children of the youth. Happy is the man that hath his quiver full of them." Psalms 127:3-5.

Chapter 12

The Embryology of the Mammal

In order to avoid unnecessary repetition, and as both the embryonic development and adult structure of the organ systems of the mammals, reptiles, and birds were covered in Chapter 11, this chapter will be devoted primarily to the early embryological stages of the placental mammal.

The placental mammals have a rather unique reproductive pattern. The eggs, instead of being laid externally, are retained within the body cavity (uterus) of the mother. Here they are protected from predators and, at the same time, are supplied with food and oxygen from the mother's blood. In addition, the embryo's metabolic wastes are carried away by the maternal blood stream. This reproductive pattern eliminates the need for a large, stored supply of yolk.

The Mammalian Egg

The mammalian egg must be considered a meiolecithal-isolecithal type. The small amount of yolk is distributed rather uniformly throughout the egg (Fig. 12-1). This equal distribution of the yolk makes it difficult to determine the position of the animal pole. The nucleus, however, is not centrally located as it undergoes its meiotic cycle. It is located in a position which is closer to the animal pole of the egg. The polar bodies are extruded at this point. These

Figure 12-1. The mammalian egg.

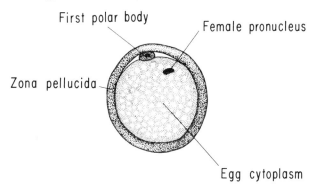

serve, therefore, as a landmark for the animal pole. It will be recalled that this is precisely the method that was utilized to determine the position of the animal pole in the other isolecithal egg type, that of the Amphioxus. Immediately surrounding the cell membrane of the egg is a noncellular layer. This noncellular layer, the zona pellucida, persists for sometime. It can be seen throughout the early cleavage stages of the mammalian egg.

Fertilization

Fertilization in the mammal is internal and usually occurs in the upper third of the oviduct. The entire process of fertilization is under the influence of both the gyno- and andro-gamones which have already been discussed in Chapter 3. The female pronucleus is located near the animal pole. If we can legitimately transfer knowledge obtained from the study of fertilization in the other vertebrate types, we might assume that the sperm copulation pathway is responsible for the establishment of the orientation of the first cleavage plane.

Cleavage

The cleavage of the mammalian egg is seen in Figure 12-2. As a result of the small amount of yolk, the cleavage pattern is the holoblastic-equal type. The first cleavage plane runs from the animal to the vegetal pole, separating the original egg into two equal blasto-

218 *The Embryology of the Mammal*

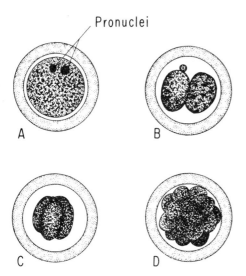

Figure 12-2. Early mammalian cleavage. (A) Fertilized egg. (B) Two-cell stage. (C) Four-cell stage. (D) The morula stage.

meres. The second cleavage plane occurs at right angles to the first, also through the animal and vegetal pole. The result of the second cleavage is four equal blastomeres. Cleavage continues until the original single egg cell has been subdivided into a number of smaller blastomeres. These blastomeres appear as a solid ball of cells referred to as the morula stage (Fig. 12-2).

Blastulation

The transformation from the solid morula to the hollow blastula occurs rather rapidly. A fluid, secreted by the cells of the morula, creates a central cavity. The blastomeres of the morula, consequently, are pushed to the periphery. This pushing outward of the cells does not occur with uniformity throughout, for in the region of the original animal pole, the cells remain multilayered, whereas around the major portion of the central cavity, the cells arrange themselves in a single layer. This creates the blastula, pictured in Figure 12-3. This part of the mammalian egg is referred to as the blastodermic vesicle. The heavier portion, near the animal pole, is referred to as the inner cell mass. This inner cell mass will ultimately give rise to the embryo proper. The central cavity is the blastocoele, whereas the single-celled layer surrounding the blastocoele is the trophoblast. This trophoblast will give rise to the extraembryonic membranes, which are so important in the development of the mammalian placenta.

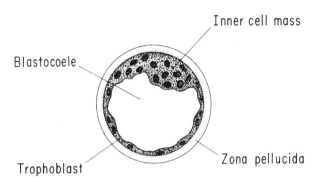

Figure 12-3. Mammalian blastula. The heavier portion near the animal pole is the inner cell mass. This will give rise to the embryo proper. The trophoblast will develop into the extraembryonic membranes.

Gastrulation and the Formation of the Primary Germ Layers

Formation of the Endoderm

Some of the ventral cells of the inner cell mass detach from their original site and begin to migrate beneath the original trophoblast. These cells, while migrating, also proliferate. Eventually, a complete layer is formed beneath the trophoblast. This is the endoderm. It can be delimited into that portion of the endoderm which lies immediately beneath the embryonic inner cell mass, the embryonic endoderm, and that portion which lies beneath the trophoblast, the extraembryonic endoderm.

Many authors have referred to the cavity within the endoderm as the primitive gut, or archenteron. However, because only part of this cavity is similar to the primitive gut and archenteron of the lower forms, I prefer to use the term "endocoele" for this cavity at this stage. The mass of cells remaining above the endoderm in the area of the original cell mass is now properly referred to as the embryonic disk. It is this embryonic disk that will give rise to the ectoderm and mesoderm of the completed embryo. An early gastrula, showing the embryonic disk, the endocoele, endoderm, blastocoele, and trophoblast, is seen in Figure 12-4. The embryonic disk, since it is to give rise to both ectoderm and mesoderm, is an example of a mesectoderm.

Figure 12-4. Formation of the mammalian gastrula. (A) Cells of the inner cell mass migrate beneath the trophoblast. (B) This new endoderm encloses an endocoelic cavity. (C) The mass of cells remaining in the original cell mass is now referred to as the embryonic disk.

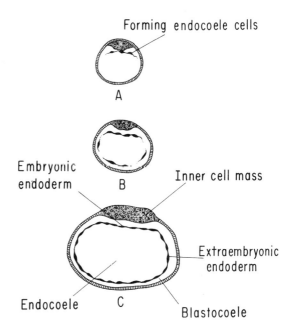

Formation of Mesoderm, Notochord, and Ectoderm

The first morphological change to occur after the formation of the endoderm is a thickening of the embryonic disk at its midcaudal extremity. This thickening is the result of a movement of cells previously described as concrescence. This concrescence occurs as the cells move toward the median line. The movement of these cells, medially as well as caudally, is depicted diagrammatically in Figure 12-5. As a result of these movements, the embryonic disk becomes oval, with the long axis of the disk coinciding with the future anterior-posterior axis of the embryo.

In a manner very similar to the formation of the mesoderm in the chick, the cells involute around the primitive ridges and migrate anteriorly and laterally between the embryonic disk and the endoderm as mesodermal sheets. Between the anteriormost extensions of this proliferating mesodermal sheet is an opening into which the developing notochord will push. This notochord comes from the same material as the mesoderm, except that this material lies in the midline of the embryonic disk. The thickened area over which the notochord material moves is referred to as Hensen's node. It is similar to the Hensen's node of the chick and the dorsal lip of the blastopore in the lower vertebrates. Once these movements have occurred, the mesoderm and notochord have been established. The remaining material on the surface of the embryo is now properly called the ectoderm. A cross section through a mammalian embryo, anterior to Hensen's node at the completion of gastrulation is seen in Figure 12-6.

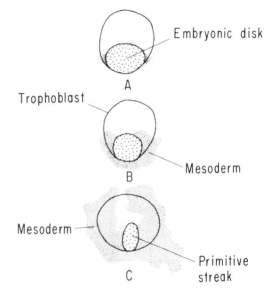

Figure 12-5. Concrescence of the epiblast cells. (A) The epiblast cells involute around the primitive ridges and migrate anteriorly and laterally between the embryonic disk and the endoderm as mesodermal sheets. (B and C) Continued formation of the mesodermal sheets.

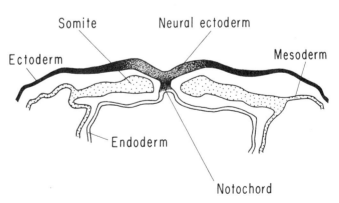

Figure 12-6. Cross section through a mammalian embryo, anterior to Hensen's node at the completion of gastrulation.

Postgastrular Mesodermal Development

The mesoderm, after involuting around the primitive streak, continues to grow in all directions between the overlying ectoderm and the underlying endoderm. This growth continues far beyond the area covered by the embryonic disk. That portion of the mesoderm which is found to lie beneath the embryonic disk is the embryonic mesoderm, whereas that portion which lies beyond is the extraembryonic mesoderm.

The mesoderm, after involution, further differentiates by splitting into an outer and inner layer. This splitting is very similar to that already seen in all the forms so far studied. The outer layer is the somatic mesoderm; the inner is the splanchnic mesoderm. The cavity between the two is the coelomic cavity. The somatic mesoderm and ectoderm form a double layer, commonly referred to as the somatopleure, whereas the splanchnic layer of mesoderm and the underlying endoderm are called, collectively, the splanchnopleure. The coelomic cavity can also be subdivided on the basis of whether or not it lies beneath the original embryonic disk. That portion which does, is the embryonic coelome. This portion of the coelome will further subdivide as development proceeds into the pericardial, pleural, and peritoneal cavities of the embryo. The extraembryonic coelome, which is found outside the limits of the embryonic disk, develops into the chorionic cavity.

Further differentiation of the mesoderm is very similar to that already seen in lower vertebrates. The mesoderm divides into basically three portions, a medial epimere or somite, an intermediate mesoderm or mesomere, and a lateral mesoderm or hypomere. The somite further differentiates into a dermatome, which gives rise to the dermis of the skin, a sclerotome, which develops into the axial skeletal structures, and a myotome, which ultimately gives rise to the striated musculature of the body. The mesomere develops into both the reproductive and excretory structures, whereas the lateral mesoderm splits into the somatic and splanchnic layers. The embryonic portions of these layers will give rise to the linings of the embryonic cavities, as well as to the heart and circulatory structures.

Postgastrular Ectodermal Development

That portion of the ectoderm which lies immediately above the notochord, after gastrulation is completed, is designated the neural ectoderm. This neural ectoderm is induced by the underlying notochord to form the neural tube, the forerunner of the nervous system.

This neural tube begins to fold up in a manner very similar to that seen in the reptile and chick. The neural folds fuse first in the midline and then proceed to close in both an anterior and a posterior direction. A dorsal view of the mammalian embryo during the closure of the neural tubes is seen in Figure 12-7. This neural tube further differentiates into the five brain regions and the spinal cord. All the structures that arise from the neural tube and the neural crest cells in the chick develop in a similar fashion in the mammal.

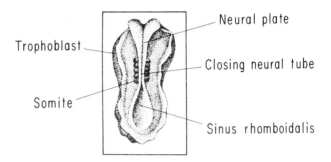

Figure 12-7. Dorsal view of the mammalian embryo during the closure of the neural tube.

Postgastrular Endodermal Development

The original flattened appearance of the embryonic disk is short-lived. A subcephalic and subcaudal folding lift the body of the embryo from the disk and establish, from the endoderm, a foregut and hindgut. That portion of the gut which is still open to the original endocoelic cavity is the midgut region. At the same time that the subcephalic and subcaudal folds are developing, lateral body folds are also helping to shape the embryo.

The derivatives of the foregut are similar to those already described for the reptile and bird. The midgut development differs somewhat in terms of the establishment of the extraembryonic membranes. These will be discussed in the section, "Formation of Extraembryonic Membranes." The hindgut development is again similar to that seen in the previous forms.

Formation of Extraembryonic Membranes

The mammal is another example of the amniote, for it possesses all of those extraembryonic membranes which have already been seen in the reptile and bird. Their embryonic significance is, however, sufficiently different to warrant a separate discussion of their formation here. Four extraembryonic membranes are present in the mammal. These are the yolk sac, amnion, chorion or serosa, and allantois. Although each of these is extremely important to the embryo during the gestation period, they are all discarded at birth. The original endodermal sac is affected by the subcaudal, subcephalic, and lateral body folds. It separates the original endodermal sac into an embryonic gut and an extraembryonic endodermal cavity. This endodermal cavity, which extends from the midgut region of the embryo, is the yolk sac. Since there is no yolk, this sac remains empty (Fig. 12-8). It is, however, associated with the same extraembryonic blood vessels, the omphalomesenteric veins which have already been seen in the lower forms. These vessels bring nutritive materials to the body of the embryo. The nutritive materials are not stored in the yolk sac but are absorbed from the uterus of the female parent by the endodermal lining of the yolk sac. In this way, the yolk sac acts in much the same manner in the mammal as it does in the reptile and bird. The lower division of the developing mesoderm, the splanchnic mesoderm, envelops the yolk sac. In reality then, it might be said that the yolk sac is composed basically of splanchnopleure. The yolk sac continues to develop and extends for some distance anterior to the head and posterior to the tail. As the other extraembryonic membranes develop, particularly the allantois, the yolk sac loses its functional importance and, as a result, decreases in size.

The amnion is created by the upfoldings of the original somatopleure. This somatopleure begins to fold up and around the embryo on all sides. Four areas are conventionally designated: the cephalic, caudal, and two lateral amniotic folds. These fuse in the region above the embryo. This area of fusion is the seroamniotic connection. The ventral connection of the amnion to the body is called the belly stalk. The formation of the amnion can be seen in Figure 12-8. The amnion, during the period of embryonic development, is filled with a watery fluid which protects the embryo from mechanical injury. This watery environment actually facilitates development by preventing contiguous parts of the embryo from adhering to one another.

In the formation of the amnion, the seroamniotic connection is formed above the embryo. Soon this dissolves, liberating the outermost portion of the somatopleure from the innermost portion. The inner portion is now the amnion. The outer portion is the chorion, or serosa. The formation of the chorion is shown in Figure 12-8. The space between the newly developed chorion and the amnion is a part of the extraembryonic coelome.

Figure 12-8. Formation of the extraembryonic membranes. (A) The splanchnic mesoderm grows down and around the endoderm to form the yolk sac. (B) The extraembryonic ectoderm, together with the underlying somatic mesoderm, begins to fold up on all sides of the embryo as the amniotic folds. (C) The amniotic folds meet and fuse above the embryo as the seroamniotic raphe. (D) The seroamniotic raphe disintegrates, creating an inner amnion and an outer serosa or chorion. (E) A ventral evagination of the hindgut, the allantois, grows into and nearly fills the chorionic cavity. It becomes intimately associated with the splanchnic mesoderm. The allantois and chorion are called, collectively, the chorioallantoic membrane.

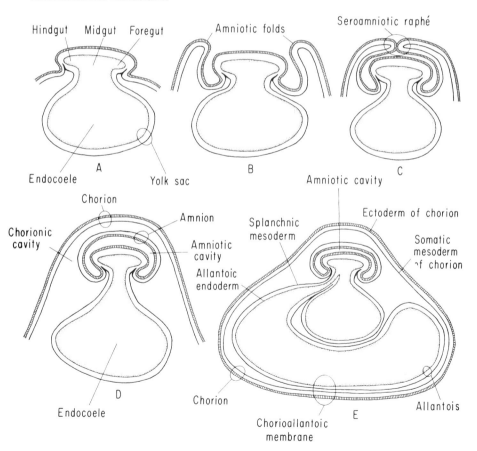

From the midventral region of the newly established hindgut, a diverticulum develops. This diverticulum is the allantois. The wall of the allantois is endoderm, but as it pushes ventrally, it acquires the splanchnic layer of mesoderm and hence becomes splanchnopleure. This allantois continues to develop until it fills nearly the entire extraembryonic coelome. It becomes intimately associated with the chorion. Together, the chorion and allantois are called the chorioallantoic membrane. Since this allantois becomes the major extraembryonic membrane in terms of metabolic exchange between the developing embryo and mother, it must be heavily vascularized. This extensive circulatory network is provided by the allantoic artery, an offshoot of the caudal end of the dorsal aorta. An embryo, after the formation of all four of these extraembryonic membranes, can be seen in Figure 12-8.

Implantation

The fertilized egg, in its early stages of development, passes through the oviduct into the uterus. Here it becomes attached to the lining of the uterus (the endometrium). This endometrial lining has been prepared for implantation by the maternal hormones. It is heavily vascularized and rather thick. Two types of implantation can be discerned. The first type is one in which the relationship between the extraembryonic membranes of the embryo and the endometrium of the mother is rather loose and at birth, the embryo, together with its membranes, is simply stripped from the uterine wall. This type is referred to as an indeciduate placenta, because the uterine lining is not shed during the birth process. The second type, the deciduate placenta, occurs where a more intimate association between the extraembryonic membranes and the uterus exists. In this type, portions of the uterus are actually destroyed and shed with the embryo at birth. The former is characteristic of mammals such as the pig, whereas the latter, or deciduate type, is found in the primates. Because the indeciduate type more clearly represents the basic association of the extraembryonic membranes with the uterine lining, it will suffice for us to dwell solely on this type of placenta.

The uterine lining is thrown into folds. The chorioallantoic membrane, composed of the ectoderm and somatic mesoderm of the chorion, and the splanchnic mesoderm and endoderm of the allantois, becomes intimately associated with these uterine folds. This brings the vascular supply of the chorioallantoic membrane into close proximity to the vascular supply of the uterine endometrium. The two blood supplies are consequently in a favorable position for diffusion to occur. Not only do nutritive supplies pass from one blood stream to another, but also gaseous exchanges between the maternal and embryonic circulation occur. This combination of chorion, allantois, and uterine endometrium is referred to as the placenta.

At birth, the amniotic sac ruptures, allowing a flow of water to precede the birth of the embryo. This is followed by the passage of the embryo itself through the uterine and vaginal canals until finally the embryo is presented at the vaginal opening. The extraembryonic membranes are born separately as an afterbirth. In an indeciduate type of placenta, there is little hemorrhaging, for the uterine wall is left intact. However, in the deciduate type of placenta, characteristic of the primates, there is some hemorrhaging, since part of the uterine wall to which the embryo was attached is lost during the birth process. Figure 12-9 diagrammatically illustrates the association of extraembryonic membranes and uterus in a typical indeciduate type of placenta.

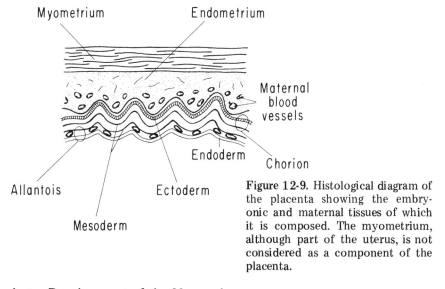

Figure 12-9. Histological diagram of the placenta showing the embryonic and maternal tissues of which it is composed. The myometrium, although part of the uterus, is not considered as a component of the placenta.

Later Development of the Mammal

The later development of the mammal is fairly similar to that in the reptiles and birds. It is not until the very late stages of development when particularly mammalian traits emerge that a difference in the embryology is noted. These later stages are beyond the scope of any elementary textbook in embryology. To illustrate the similarity of development of the mammal, the reptile, and the bird, Figure 12-10 represents a 10-mm pig embryo with most of the major structures depicted. It can be seen from this figure that the developing structures are very similar to their counterparts in the lower forms. The beauty of vertebrate embryology is that even the names of most of the structures remain the same from one form to another.

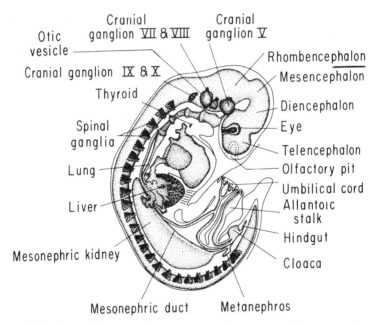

Figure 12-10. Ten millimeter pig embryo showing most of the major structures which have been discussed.

"Rejoice in the Lord always: and again I say, Rejoice."
Phillipians 4:4.

Chapter **13**

In Retrospect

There is a great deal of truth in the old adage that a little knowledge is a dangerous thing. There is danger also in writing a textbook of this type, for if it gives to the reader a sense of embryological proficiency, it has missed its point. The mere reading of this textbook or the taking of an elementary course in embryology does not entitle one to think he is an embryologist. This book has purposely been written on an elementary level, with the belief that if these basic embryological facts can be mastered by the student, it will be easier for him to go on to more advanced studies in this area. The danger of omission is always present. It is hoped that the omissions in this book have not led to erroneous embryological concepts.

In addition to presenting basic embryological facts, this book should have raised many important questions in the mind of the reader. The student would be justified in asking why the structures presented in this book are so similar in so many widely divergent animals. He might also question the relationship of the development of a particular structure to the animal's ecological environment, or he might wonder what forces are at work to cause a specific portion of the neural tube to evaginate as an eye vesicle. These, and many other questions, are legitimate and should be encouraged. For the most part they are still unanswered. It will take young, fertile minds to give us the answers to these many perplexing problems. The student, therefore, is encouraged to read extensively in embryological literature. Many good books are available describing the specific development of the individual animals discussed in this book. In addition, journal articles should be referred to. One can never read too much in an area to increase one's competency.

Finally, this book should have given to you an appreciation of the complexity of the embryonic process. In these times of high-powered nuclear bombs and almost human, room-sized computers, the developing embryo might seem insignificant. To the true student of embryology, however, man-made devices will never surpass the complexity shown by the development that occurs over a 24-hour period in an embryo.

For some students this course will be their last in embryology; for others it will be only the beginning. It is the fervent hope of the author that this book has helped to awaken in some students a desire to go on with their embryological training. If this end has been met, the existence of this book is justified.

Index

Abducens, 135
Acousticofacialis ganglion, 185
Acrosome, 10, 11
Adrenal gland, 140
Adrenal medulla, 114
Afferent dendrite, 132
Afferent renal vein, 151
Afferent vessel, 146
Afterbirth, 227
Air sacs, 210
Albumen, 157, 170
Alecithal eggs, 13, 14
Allantoic arteries, 205, 207, 226
Allantoic cavity, 193
Allantoic endoderm, 225
Allantoic sac, 193
Allantoic stalk, 195, 228
Allantoic vein, 195, 207
Allantois, 23, 192, 193, 194, 201, 207, 224, 225, 226, 227
Amnion, 23, 192, 193, 224, 225
Amniote, 192, 223
Amniotic cavity, 23, 182, 193, 195, 225
Amniotic fold, 180, 192, 224, 225
Amniotic sac, 227
Amniotic umbilicus, 192
Amphibian blastula, fate map of, 108
Amphibian egg, cleavage in, 106, 107
Amphibian egg, fertilization of, 104
Amphibian egg, morphology, 104
Amphibian gastrulation, 108, 109
Amphibian heart, development of, 147
Amphibian sperm, 11
Amphioxus, blastula formation, 74, 75, 76
Amphioxus, cleavage pattern, 72, 73
Amphioxus, development, 68
Amphioxus, development of endoderm, 82
Amphioxus egg, 68, 69, 74
Amphioxus, fertilization, 70
Amphioxus, mesodermal development, 80
Amphioxus sperm, 11
Amplexus, 104
Anal opening, 84
Anal plate, 84, 101, 186

Anamniotes, 192
Anaphase, 13, 14, 15, 18
Anastamoses, 200
Androgamones, 27
Animal hemisphere, 74, 104
Animal micromeres, 107
Animal pole, 69
Animal-vegetal axis, 72
Animal-vegetal polarity, 39
Anterior abdominal air sac, 209
Anterior appendage bud, 201
Anterior cardinal veins, 149, 150, 179, 181, 191, 195, 205, 207
Anterior choroid plexus, 130, 131
Anterior intestinal portal, 165, 166, 177, 184, 186, 199
Anterior liver diverticulum, 200
Anterior neuropore, 55, 57, 79, 80, 116, 164, 176, 178
Anterior subcardinal extension, 151
Anterior vena cava, 211
Anterior vitelline vein, 191
Aorta, 179
Aortic arches, 147, 148, 182, 186, 189, 190, 194, 195, 204, 206
Aortic arches, development of, 149
Aortic arches, differentiation of, 148
Aortic arches, fate of, 206
Aqueduct of Sylvius, 130, 197
Archenteric canal, 162, 163, 164, 165
Archenteron, 40, 47, 48, 54, 56, 57, 78, 79, 80, 109, 162, 174, 220
Arcualia, 143, 144, 202
Arcualia, derivatives of, 144
Area opaca, 41, 160, 161, 171, 172
Area pellucida, 41, 160, 161, 162, 171, 172
Arteries, 147
Artificial parthenogenesis, 22
Association axon, 135
Association dendrite, 135
Association nerve cell body, 135
Association neurons, 198
Atrioventricular canal, 204
Atrium, 146, 147, 182, 189, 194, 203, 204, 207, 211
Attached sperm, 21
Auditory nerve, 139
Auditory placodes, 117, 139, 185

Auditory sac, 139
Auditory vesicle, 182, 194, 195, 201
Autonomic ganglia, 114
Autonomic nervous system, 136
Aves sperm, 11
Axial filament, 11, 12
Axillary vein, 211
Axons, 133

Balfour's law, 32
Barbs, 208
Basidorsal, 143, 144
Basiventral, 143, 144
Belly stalk, 224
Bicuspid valve, 204
Bile duct, 119, 141, 142
Blastocoele, 39, 40, 42, 43, 46, 72, 73, 107, 109, 172, 218, 219, 220
Blastocoele, primary, 89, 160
Blastocoele, secondary, 161
Blastoderm, 39, 89, 90, 91, 93, 96, 159, 160
Blastodermic vesicle, 218
Blastodisk, 41, 172
Blastomeres, 29
Blastoporal canal, 162
Blastopore, 44, 45, 46, 47, 56, 57, 77, 93
Blastopore, development of, 113
Blastopore, dorsal lip of, 45, 109
Blastopore, lateral lips of, 45
Blastopore, lip, 44
Blastopore, ventral lip of, 45
Blastula, 28, 39, 47, 107, 108, 218
Blastula, of the amphioxus, 75, 76
Blastula, formation, in amphioxus, 74
Blastula, of an isolecithal egg, 40
Blastula, of a moderately telolecithal egg, 40
Blastula, reptilian, fate map of, 161
Blastula, in teleost, 88
Blastulation, 39
Blastulation, in a heavily telolecithal egg, 40
Blastulation, in an isolecithal egg, 39
Blastulation, mammal, 218
Blastulation, in a moderately telo-lecithal egg, 40
Blastulation, reptile, 160
Blood islands, 178
Blood vessels, 179, 180, 208
Blood vessels, development, 205
Blood vessels, maternal, 227
Body, 126
Body ectoderm, 108, 126, 132

Body shape, formation of, reptiles, 164
Body shape, teleost, 97
Brachial nerve, 133
Brachial plexus, 133
Brachiocephalic vein, 154
Brain, 165, 196, 223
Brain, amphibian, development of, 129
Brain, ventricle, 130, 131, 196, 197
Branchial apparatus, 83, 84
Branchial arches, 112, 125
Branchial cleft, 112, 114, 123
Branchial cleft, development of, 112
Bronchii, 141, 186, 209
Bulbous arteriosus, 146, 147, 190

Carapace, 176
Cardinal veins, anterior, 191
Cardinal veins, posterior, 191
Cardinal veins, system, 149
Carotid glands, 140, 141
Carpals, 207, 208
Cartilage, 208
Cartilagenous bone, 208
Caudal amniotic fold, 224
Caudal artery, 123, 211
Caudal flexure, 165, 196, 201
Caudal vein, 123, 211
Central cells, 171
Central nervous system, 110, 198
Central periblast, 159, 160
Centrolecithal, 86, 88
Centrosomes, 11
Centrum, 144
Cephalic amniotic fold, 224
Cephalic flexure, 165
Cerebellar hemisphere, 197
Cerebellum, 130, 131
Cerebral aqueduct, 130, 131, 197
Cerebral hemisphere, 129, 184, 196
Cervical air sac, 209
Cervical flexure, 165, 166, 180, 196, 201
Chalaza, 170
Chemical embryologists, 4
Chemistry of fertilization, 26, 27
Chiasmata, 16, 18
Chick, blastoderm, fate map of, 173
Chick, embryology of, 169
Choanae, 118, 140
Chondroblasts, 203, 207, 208
Chondrocranium, 208
Chondrocytes, 203, 208
Chordal cells, 77, 78
Chordamesoderm, 53, 55, 58, 109

Chordaneural crescent, 76, 78
Chorioallantoic membrane, 194, 209, 226, 227
Chorion, 23, 192, 193, 194, 224, 225, 226, 227
Chorionic cavity, 194, 222, 225
Choroid fissure, 138
Choroid plexus, anterior, 130
Choroid plexus, posterior, 131, 197
Chromatids, 16
Chromatophores, 114, 140
Chromonemata, 16
Chromosomal aberrations, 16, 17
Chromosome, homologous, 13
Chyme, 210
Circular muscles, 210
Circulatory pattern, teleosts, 101, 102
Circulatory system, 211
Clavicular air sac, 209
Cleavage, 28, 29, 31, 35, 37
Cleavage, amphibian, 106, 107
Cleavage, amphioxus, 72, 73
Cleavage, chick, 170, 171
Cleavage, compared with mitosis, 30
Cleavage, discoidal, 33, 36
Cleavage, holoblastic equal, 32, 33
Cleavage, holoblastic unequal, 32, 34
Cleavage, mammal, 217
Cleavage, meroblastic, 33
Cleavage, reptile, 159
Cleavage, rules of, 32
Cleavage, significance, 36
Cleavage, teleost, 88, 89
Cleidoic egg, 23, 24, 157, 170, 191, 192
Cloaca, 60, 118, 145, 186, 188, 203, 211, 214, 228
Cloacal opening, 101
Cloacal plate, 101, 119, 186
Club-shaped gland, 83
Cochlea, 139
Coelome, 53, 59, 62, 82, 97, 121, 145, 177, 179
Coelomic cavity, 28, 60, 81, 96, 121, 177
Common bile duct, 200
Common cardinal veins, 150, 190, 191, 207
Common carotid artery, 149, 205, 211
Common iliac artery, 154
Common iliac vein, 211
Comparative embryology, 4
Concrescence, 220
Concrescence, of the epiblast cells, 220

Cone, fertilization, 24, 25
Cones, 138
Conus, arteriosus, 147, 190, 203
Convergence, 50
Copulation path, 26, 205
Cornea, 138
Corpora bigemina, 130
Corpora quadrigemina, 197
Cortex, 21
Cortical cytoplasm, 69, 71, 105
Corticoplasma, 71
Cranial flexure, 180, 195, 201
Cranial ganglia, 114, 185, 198, 228
Cranial nerves, 135, 139, 198
Cranial nerves, distribution, 136
Cranial nerves, origin, 136
Craniosacral system, 136
Crop, 210
Crura cerebri, 130, 131, 197
Cushion septum, 204
Cutous plate, 120
Cystic duct, 200
Cytoplasm, 12, 87
Cytoplasm, cortical, 69, 71, 105

Deciduate placenta, 226, 227
Delamination, 41, 42, 161, 171
Deletion, 16, 17
Dendrite, 133
Dermal bone, 208
Dermatocranium, 208
Dermatome, 82, 120, 143, 186, 202, 222
Dermis, 120, 222
Descriptive embryology, 3
Deutoplasm, 32, 170
Developmental physiology, 4
Diad, 15, 18
Diakinesis, 15, 18
Diencephalon, 117, 123, 129, 137, 138, 181, 182, 183, 184, 194, 195, 196, 197, 201, 228
Diencoele, 130, 182, 197
Differential growth, 207
Digestive tract, 209
Diplotene, 15, 17
Discoidal cleavage, 33, 36, 170
Distal centrosome, 12
Diverticulum, left, 82
Diverticulum, right, 82
Dorsal aorta, 101, 102, 147, 148, 149, 179, 182, 190, 191, 195, 205, 207, 211, 226
Dorsal diverticulum, 82
Dorsal fin, 123
Dorsal ganglion, 132, 133, 195

233

Dorsal horns, 195
Dorsal lip, 45, 46, 76, 77, 78, 91, 93, 95, 108, 109
Dorsal mesentery, 188
Dorsal mesocardium, 179, 182
Dorsal root, 179
Dorsal spinal ganglia, 198
Dorsal thickening, 116
Duct, of Bottalus, 205
Duct, of Cuvier, 150, 151, 205
Ductus arteriosus, 148, 149, 205
Ductus choledochus, 200, 203, 210
Ductus deferens, 215
Ductus venosus, 190, 200
Duodenum, 203, 210

Ear, 139
Ear, amphibian, development of, 139
Ear, ossicles, 209
Ectoderm, 28, 43, 44, 48, 49, 54, 57, 58, 60, 61, 62, 72, 75, 76, 82, 84, 90, 92, 97, 100, 101, 161, 162, 165, 174, 179, 192, 194, 220, 221, 227
Ectoderm, of chorion, 225
Ectoderm, development, amphibian, 114
Ectoderm, development, chick, 183
Ectoderm, development, mammal, 222
Ectoderm, of head, 138, 179
Ectoderm, sheet, 80
Efferent axon, 132
Efferent renal vein, 150
Efferent vessel, 145
Egg, 3, 6, 8, 9, 10, 24
Egg, amphioxus, 68, 69
Egg, cleidoic, 23, 24
Egg, cytoplasm, 216
Egg, function, 8
Egg, isolecithal, 34
Egg, membrane, 170
Egg, morphology, 12
Egg, morphology, chick, 170
Egg, morphology, reptile, 157
Egg, morphology, teleost, 86
Egg, nucleus, 170
Egg, orientation, amphioxus, 74
Egg, rotation, 54
Egg, shell, 170
Egg, symmetry, 71
Ejaculatory duct, 215
Embryo, 23
Embryology, 2, 3
Embryology, of the chick, 169
Embryology, of the mammals, 216
Embryology, of the reptiles, 156

Embryology, of the teleosts, 85
Embryonic axis, origin of, 93
Embryonic coelome, 179
Embryonic disk, 220, 221
Embryonic periblast, 89
Embryonic shield, 91, 162
Endocardial tube, 147, 188
Endocardium, 179, 182
Endocoele, 219, 220, 223, 225
Endocoele, cells, 219
Endocrine glands, 140
Endoderm, 28, 43, 48, 49, 57, 58, 60, 61, 62, 72, 75, 82, 90, 92, 93, 95, 101, 108, 162, 165, 174, 177, 179, 193, 195, 221, 227
Endoderm, development of, amphibian, 118
Endoderm, development of, amphioxus, 82
Endoderm, development of, chick, 186, 198
Endoderm, development of, mammal, 223
Endoderm, embryonic, 219
Endoderm, extraembryonic, 219
Endoderm, formation of, 219
Endometrium, 226, 227
Endostyle, 83
Endothelial tube, 146
End piece of sperm, 12
Enteron, 95
Epiblast, 41, 42, 43, 61, 62, 92, 161, 162, 163, 171, 172, 173
Epiblast, cells, concrescence of, 220
Epiboly, 47, 109
Epidermis, 121, 123
Epididymal duct, 7, 8
Epididymis, 7
Epimere, 59, 60, 61, 62, 120, 121, 177, 179, 186, 222
Epimyocardium, 147, 177, 179, 180, 188
Epiphysis, 130, 184, 194, 196, 201
Equatorial plate, 13
Esophagus, 118, 141, 142, 199, 210
Estrogen, 9
Excretory system, 145, 186, 213
Experimental embryology, 4
External carotid arch, 206
External carotid artery, 148, 149, 154, 205, 207
External fertilization, 23
External gills, 123, 125, 126
External jugular vein, 151, 205
External nares, 123, 140
Extraembryonic coelome, 179, 182, 193

Extraembryonic membrane, 23, 24, 191, 218, 227
Extraembryonic membrane, development of, 193, 223, 225
Extraembryonic periblast, 89, 91
Eye, 116, 137, 166, 198, 228
Eye, development of, 138
Eye, teleost, 99

Facial nerve, 136
Fat body, 154
Fate map, amphibian blastula, 108
Fate map, chick blastula, 173
Fate map, reptilian blastula, 161
Fate map, teleost blastula, 90
Feather germ, 208
Feather germ, development of, 208
Feather pulp, 208
Feathers, formation of, 208
Female pronucleus, 21, 26, 69, 87, 105, 216
Female reproductive system, development of, 214
Female sex hormone, 9
Femoral artery, 154, 211
Femoral vein, 154, 211
Fertilization, 6, 20
Fertilization, amphibian, 104
Fertilization, amphioxus, 70
Fertilization, chemistry of, 26
Fertilization, chick, 170
Fertilization, cone, 21, 24, 25
Fertilization, external, 23
Fertilization, importance of, 24
Fertilization, internal, 23
Fertilization, mammal, 217
Fertilization, membrane, 21, 22, 25
Fertilization, reptile, 157
Fertilization, teleost, 86
First polar body, 9
Flexion, 165, 195, 201
Flexure, 180, 194
Follicular cells, 9, 10
Foramen of Munro, 129, 131, 196
Foregut, 60, 100, 102, 118, 140, 152, 165, 166, 177, 178, 179, 186, 200, 202, 223, 225
Foregut, derivatives, 141
Formation, of eggs, 8
Formation, of gametes, 6
Formation, of gray crescent, 105
Formation, of mesoderm, 51
Formation, of neural tube, 57, 80
Formation, of primary germ layers, mammal, 218
Formation, of primitive streak, 49

Formation, of sperm, 6
Frog heart, 146
Fusion nucleus, 87

Gall bladder, 118, 119, 141, 142, 186, 200, 203
Gametes, 3, 14
Gametes, formation, 6
Gametogenesis, 3, 6
Gamones, 27
Ganglion, 180, 185, 194, 195
Gastrocoele, 47
Gastrula, 28, 45
Gastrulation, 28, 43, 76
Gastrulation, amphibian, 108, 109
Gastrulation, chick, 171
Gastrulation, heavily telolecithal egg, 48, 50
Gastrulation, isolecithal egg, 44
Gastrulation, mammal, 218
Gastrulation, moderately telolecithal egg, 45, 46
Gastrulation, movements, teleost, 91
Gastrulation, reptile, 162
Gastrulation, teleost, 90, 91
Genital ridges, 154
Germ, cells, origin of, 215
Germ, ring, 91, 92, 93
Germ, wall, 159
Germinal disk, 13, 14, 36, 170
Gill, arch development, chick, 186
Gill, chamber, 126
Gill, placode, 125
Gill, plate, 111, 118, 119
Gill, slits, 84, 180
Gizzard, 199, 210
Gizzard, plates, 210
Glomerulus, 145
Glossopharyngeal, 136
Glottis, 141, 186, 202, 209
Golgi apparatus, 10
Gonadal ridge, 214
Gonads, 154, 214
Graafian follicle, 8, 9, 10
Gray crescent, 106
Gray crescent, formation of, 105
Gray matter, 132, 197
Gut, 82, 97, 193, 195
Gut cavity, 60, 95, 96, 97, 109
Gynogamone, 27

Head ectoderm, 138, 139, 165
Head kidney, 145
Head process, 51, 60, 173
Head process, formation of, 174
Head, sperm, 11

Heart, 102, 147, 148, 149, 150, 180, 188, 189, 206
Heart, amphibian, development of, 147
Heart, development, 146, 166, 188, 203
Heart, muscle, 146
Heavily telolecithal egg, 13, 14
Heavily telolecithal egg, neurulation in, 62
Hemal arch, 144
Hemal spine, 144
Hensen's node, 49, 50, 172, 173, 174, 175, 176, 221
Hepatic duct, 141, 200
Hepatic portal system, 152
Hepatic portal system, development of, 152
Hepatic portal veins, 152, 153, 154, 211
Hepatic sinusoids, 152
Hepatic vein, 152, 153, 154
Hertwig's rules, 32
Heterogonic growth, 207
Hindgut, 60, 100, 101, 102, 118, 140, 142, 165, 166, 186, 223, 225, 228
Holoblastic equal cleavage, 32, 33
Holoblastic unequal cleavage, 32, 34
Homologous chromosome, 13
Humerus, 207, 208
Hyaline cartilage, 203
Hyaloplasm, 71
Hyoid arch, 112, 113, 119, 125, 186
Hyoid bone, 209
Hyomandibular cleft, 112, 119, 194
Hyomandibular groove, 119
Hypoblast, 41, 42, 43, 92, 95, 161, 162, 163, 171, 173
Hypoblast, origin of, 41
Hypoglossal nerve, 133
Hypomere, 59, 61, 62, 120, 121, 145, 177, 186, 222
Hypophysis, 116, 130, 185

Iliac artery, 211
Iliac artery, external, 207
Implantation, mammal, 226
Indeciduate placenta, 226, 227
Induction, 198
Infiltration, 43, 161, 171
Infundibulum, 116, 117, 130, 179, 184, 185, 197
Inner cell mass, 218, 219
Interatrial septum, 204
Interdorsal, 143, 144

Intermediate mesoderm, 58, 59, 61, 142, 144, 177, 179, 222
Internal carotid arch, 206
Internal carotid artery, 148, 149, 154, 205, 207
Internal fertilization, 23
Internal iliac vein, 211
Internal jugular vein, 151, 205, 207
Internal nares, 140
Interphase, 14
Interventral, 143, 144
Interventricular septum, 167, 204
Intestine, 211
Intraembryonic coelome, 181
Invagination, 44
Inversions, 17
Involution, 42, 45, 47, 161, 171, 173
Isolecithal, 14
Isolecithal, blastula of, 40
Isolecithal egg, 13, 34
Isolecithal, mesoderm formation, 53
Isolecithal, neurula, 57
Isolecithal, neurulation in, 55
Isolecithal, notochord formation, 53

Jugular veins, 205, 211

Kidney, 151

Large intestine, 118
Laryngeal cartilages, 209
Laryngotracheal groove, 186, 195, 199, 202, 209
Larynx, 141, 186, 199, 202, 209
Lateral amniotic fold, 224
Lateral body fold, 98, 224
Lateral lip, 45, 77
Lateral mesoderm, 58, 59, 61, 177, 222
Lateral plate mesoderm, 121, 142
Left atrium, 146, 204
Left diverticulum, 82
Left ventricle, 204
Legs, 196
Lens, 125, 131, 137, 138, 182, 198
Lens primordium, 138
Leptotene, 15, 16
Limb buds, 128, 167, 196, 201, 207
Liver, 60, 118, 119, 141, 142, 152, 153, 154, 186, 195, 199, 200, 203, 210, 211, 228
Liver, diverticulum, 152
Liver, lobe, 200
Longitudinal muscles, 210
Lumbar flexure, 196, 201
Lumbosacral plexus, 134

Lumen, 210
Lung buds, 199, 203
Lungs, 141, 142, 148, 149, 154, 186, 206, 209, 211, 228

Macromeres, 35, 73, 74, 106, 107
Male pronucleus, 21, 69
Male reproductive system, development of, 214
Mammal, egg, 216
Mammal, embryology of, 216
Mammal, sperm, 11
Mandibular arches, 111, 112, 113, 119, 182, 186, 195
Marginal cells, 171
Maxillary process, 195
Medulla oblongata, 130, 131
Medullary canal, 57
Medullary plate, 55, 56, 57, 59, 62, 79, 80, 176
Meiolecithal eggs, 13
Meiosis, 13, 14
Meiosis, significance, 19
Membrane, extraembryonic, 23, 24
Membrane, extraembryonic, fertilization, 25
Membrane, extraembryonic, vitelline, 25
Meroblastic cleavage, 33, 36, 170
Mesencephalon, 60, 99, 115, 116, 130, 178, 180, 183, 184, 194, 195, 201, 228
Mesenchymal cells, 213
Mesenchyme, 179
Mesencoele, 130, 179
Mesendoderm, 44, 48, 54, 58, 163
Mesoderm, 28, 43, 49, 50, 58, 61, 72, 75, 76, 90, 92, 93, 108, 109, 161, 163, 165, 173, 174, 220, 221, 227
Mesoderm, crescent, 71, 72, 73, 74, 75
Mesoderm, development, amphibian, 120, 142
Mesoderm, development, amphioxus, 80
Mesoderm, development, chick, 186
Mesoderm, development, mammal, 221
Mesoderm, differentiation, teleost, 96
Mesoderm, formation of, 51, 174
Mesoderm, formation of, heavily telolecithal egg, 60
Mesoderm, formation of, isolecithal egg, 53

Mesoderm, formation of, mammal, 220
Mesoderm, formation of, moderately telolecithal egg, 58
Mesoderm, intermediate, 58, 59
Mesoderm, somatic, 53, 58, 59, 61, 81, 96
Mesoderm, splanchnic, 53, 58, 59, 61, 81, 96
Mesomere, 59, 60, 61, 62, 120, 121, 144, 145, 177, 186, 222
Mesonephric cup, 145
Mesonephric duct, 145, 155, 182, 203, 213, 214, 215, 228
Mesonephric kidney, 145, 213, 214, 228
Mesonephric tubule, 145, 155, 182, 214
Mesonephros, 145, 150, 186
Metacarpals, 207, 208
Metamorphosis, 7, 8, 9, 122, 127
Metanephric duct, 213, 214
Metanephric kidney, 213, 214
Metanephric kidney, development of, 213
Metanephros, 228
Metaphase I, 13, 14, 15, 18
Metencephalon, 183, 184, 194, 197, 201
Micromeres, 35, 72, 73, 74, 106, 107
Micropyle, 87
Midbrain, 184
Middle piece of sperm, 12
Midgut, 100, 118, 121, 140, 142, 166, 177, 179, 186, 193, 203
Midsagittal plane, 70
Migrating neural crest cells, 132
Mitochondria, 11, 12
Mitosis, compared with cleavage, 30
Mitral valve, 204
Moderately telolecithal egg, 13, 14
Moderately telolecithal egg, blastula of, 40
Moderately telolecithal egg, neurulation of, 59
Monospermy, 21
Morphology, of egg, amphibian, 104
Morphology, of the sperm, 10
Morula, 34, 218
Motor axon, 135
Motor neurons, 198
Mouth, 83, 100, 118, 125, 142
Mucosa, 209, 210
Mucous gland, 111, 112, 114
Mucous gland, development, 112
Mullerian duct, 155, 215

Munro, foramen of, 129, 131, 196
Muscle, circular, 210
Muscle, longitudinal, 210
Mutagenic agents, 17
Myelencephalon, 131, 182, 183, 184, 194, 197, 198
Myelocoele, 181, 197
Myocardium, 146
Myocoele, 82, 120, 143, 186
Myometrium, 227
Myotome, 82, 113, 114, 120, 122, 123, 186, 202, 222

Nasal chambers, 139
Nasal pits, 198
Nasal placodes, 117, 198
Nephrocoele, 121
Nephrotome, 120, 121
Nervous ganglia, 148
Nervous system, 222
Nervous system, chick, 178, 196
Neural arch, 144
Neural canal, 57, 79
Neural cells, 115, 120, 132, 133, 140, 177, 185, 223
Neural crest, 114, 179
Neural crest, cells, 115, 120, 132, 133, 140, 177, 185, 223
Neural ectoderm, 93, 108, 110, 115, 132, 161, 173, 221, 222
Neural folds, 59, 164, 165, 176, 177, 223
Neural groove, 110, 115
Neural plate, 76, 77, 78, 94, 110, 164, 165, 223
Neural ridges, 55, 56, 57, 110, 115
Neural spine, 144
Neural tube, 28, 55, 57, 59, 60, 62, 79, 80, 82, 90, 94, 97, 111, 114, 115, 120, 121, 123, 144, 165, 176, 177, 178, 185, 193, 195, 222, 223
Neural tube, area, 93
Neural tube, development of, 115
Neural tube, development of, teleost, 99
Neural tube, formation of, 57
Neural tube, formation of, teleost, 95
Neurenteric canal, 55, 57, 80, 164, 165
Neuroblasts, 132, 133, 135, 184
Neurocoele, 57, 59, 79, 80, 82, 94, 95, 96, 97, 120, 131, 132, 164, 177

Neurocranium, 208
Neuroglial cells, 132, 185
Neuromeres, 183, 184
Neurons, 185
Neurula, 28, 60
Neurula, amphibian, 111
Neurulation, 52
Neurulation, chick, 176
Neurulation, heavily telolecithal egg, 62
Neurulation, isolecithal egg, 55, 57
Neurulation, moderately telolecithal egg, 59
Neurulation, reptile, 164
Neurulation, teleost, 94
Nictitating membrane, 169
Nonnervous ectoderm, 115, 132, 173
Notochord, 28, 50, 51, 57, 58, 59, 60, 61, 62, 76, 78, 79, 81, 82, 83, 90, 92, 93, 95, 96, 97, 108, 109, 110, 120, 121, 123, 143, 161, 163, 164, 165, 173, 174, 177, 221
Notochord, canal, 162
Notochord, cells, 78
Notochord, formation, 174
Notochord, formation, isolecithal egg, 53
Notochord, mammal, formation of, 220
Notochord, plate, 78
Nucleoplasmic ratio, 30
Nucleus, 10, 12

Oculomotor nerve, 136
Olfactory lobe, 129, 131
Olfactory nerves, 129, 131, 136, 140, 198
Olfactory pit, 123, 125, 194, 198, 201, 228
Olfactory placodes, 139
Omphalomesenteric arteries, 102, 180, 190, 191, 194, 195, 207
Omphalomesenteric capillaries, 102
Omphalomesenteric veins, 102, 147, 152, 178, 179, 182, 188, 189, 190, 191, 194, 195, 201, 207, 224
Ontogeny, 2, 3
Oögenesis, 6, 9
Oögonia, 9, 155
Oöplasm, 69
Oötid, 9, 15
Opercular membrane, 128
Operculum, 125, 126
Optic chiasma, 130, 131

Optic cup, 131, 137, 138, 180, 198
Optic lobes, 130, 131
Optic nerves, 130, 136, 139
Opticoele, 179
Optic primordia, 116, 117
Optic recess, 130, 184
Optic stalk, 116, 117, 137, 138
Optic vesicle, 99, 111, 112, 114, 122, 130, 138, 178, 184, 185
Optic vesicle, development of, 117
Oral cavity, 83, 152
Oral evagination, 100
Oral plate, 100, 118, 186, 198
Oral sucker, 111, 112
Osteoblasts, 208
Osteocytes, 208
Ostium, 8, 9, 214
Otic pit, 125
Otic placode, 139
Otic vesicle, 139, 180, 185, 198, 228
Otocyst, 185
Ova, 13
Ovary (ies), 8, 155, 214, 215
Oviduct, 8, 214, 226
Oviparous, 23
Ovoviviparous, 23
Ovulation, 8, 9

Pachytene, 15, 16
Pancreas, 118, 142, 203
Pancreatic diverticulum, 141, 203
Pancreatic duct, 142, 210
Parasympathetic ganglia, 136, 137
Parasympathetic nervous system, 136
Parasympathetic postganglionic neuron, 137
Parasympathetic preganglionic neuron, 137
Parathyroid glands, 140, 141
Pars distalis, 116, 185
Pars intermedia, 116, 185
Pars nervosa, 116, 185
Pars tuberalis, 116, 185
Parthenogenesis, 22
Pectoral artery, 211
Pectoral girdle, 208
Pectoral vein, 211
Pelvic vein, 154
Penetration path, 26, 69, 105
Perch sperm, 11
Periblast, 89, 90, 93, 96, 97, 159
Periblast, central, 159
Periblast, peripheral, 159, 160, 161
Periblast, reptile, formation of, 159
Pericardial cavity, 146, 222

Pericardial sac, 146
Pericardial space, 189
Peripheral nervous matter, 198
Peripheral nervous system, 133
Peripheral periblast, 159, 160, 161
Peritoneal cavity, 222
Perivitelline space, 25, 69, 87, 97, 98
Permanent subclavian, 206
Phlanges, 207, 208
Pharyngeal pouch, 118, 140, 186
Pharyngeal region, 119
Pharynx, 84, 118, 141, 142, 182, 202
Pigeon milk, 210
Pigment cells, 114, 140
Pigmented coat, 137
Pigmented cortical plasm, 105
Pigment layer, 138, 185
Pineal gland, 130, 131, 184, 196
Pituitary gland, 117, 130, 185, 197, 202
Placenta, 23, 227, 228
Placenta, deciduate, 226, 227
Placenta, indeciduate, 226, 227
Planes of symmetry, 70
Plasma membrane, 11, 12
Plastron, 167
Pleural cavity, 199, 222
Pleuroperitoneal cavity, 146
Point of fusion, 80
Point mutation, 17
Poison effect, 17
Polar body, first, 9, 73, 78, 216
Polarity, animal-vegetal, 39
Polylecithal eggs, 13
Polyspermy, 21, 157
Pons varolii, 197
Postanal gut, 100, 101
Postcardinal vein, 151
Postcaval vein, 151, 154
Postcaval vein, development of, 151
Postcaval vena cava, 151
Posterior abdominal air sac, 209
Posterior appendage bud, 201
Posterior cardinal vein, 145, 149, 150, 182, 191, 195, 207
Posterior choroid plexus, 131
Posterior intestinal portal, 165, 166, 186, 201
Posterior liver diverticulum, 200
Posterior mesenteric vein, 211
Posterior vena cava, 151, 206, 211
Posterior vena cava, development of, 150
Posterior vitelline vein, 191
Precartilage, 207
Prechordal area, 90

Prechordal material, 91, 173
Prechordal plate, 161
Prechordal tissue, 92
Preoral gut, 100, 186, 199
Preoral pit, 82
Presumptive lens area, 138
Presumptive optic area, 117
Primary blastocoele, 89, 160
Primary oöcytes, 9
Primary spermatocyte, 7, 8
Primitive fold, 49, 50, 172
Primitive groove, 49, 50
Primitive gut, 109, 220
Primitive gut, development of, 99
Primitive gut, formation of, teleost, 101
Primitive pit, 49, 50, 172
Primitive plate, 162
Primitive ridges, 173
Primitive streak, 49, 61, 113, 172, 173, 174, 175, 176, 178, 220
Primitive streak, formation of, 49
Primitive streak, regression of, 176
Primordial germ cells, 7, 8, 9
Proammion, 178
Proctodeum, 100, 101, 113, 119, 126, 140, 187
Pronephric duct, 145, 186, 188
Pronephric kidney, 121, 145
Pronephric tubules, 145, 186, 188
Pronephros, 7, 114, 124, 145, 155, 186, 188
Pronuclei, 20, 217
Prophase I, 13, 14, 15, 16, 18
Prosencephalon, 60, 98, 99, 115, 116, 129, 178, 183
Prosencoele, 179
Proventriculus, 199, 210
Proximal centrosome, 11, 12
Pulmocutaneous artery, 154
Pulmonary arch, 148, 206
Pulmonary artery, 148, 149, 203, 205, 209, 211
Pulmonary vein, 154, 211
Pupil, 137, 185

Radius, 207, 208
Rathke's pocket, 116, 117, 185, 197
Recessus opticus, 184
Rectum, 118
Reflex arc, 135
Renal cortex, 213
Renal medulla, 213
Renal pelvis, 213, 214
Renal portal system, 151
Renal portal vein, 211

Renal tubules, 214
Renal vein, 151, 153, 154
Reproductive system, 153, 154, 213
Reptile, embryology of, 156
Respiratory system, 141, 209
Rete testis, 7, 8
Retina, 137, 138, 185, 195
Rhombencephalon, 60, 99, 115, 116, 129, 130, 139, 178, 180, 183, 228
Rhombencoele, 130
Rods, 138
Rotation, egg, 54

Saccule, 139
Sach's rules, 32
Saunders' ridge, 196
Sciatic artery, 154, 205, 211
Sciatic plexus, 134
Sciatic vein, 211
Sclerotomal cells, 120, 144, 186, 202
Sclerotomal sheath, 143
Sclerotome, 60, 82, 120, 121, 143, 222
Scrotal sac, 7
Second polar body, 9
Secondary blastocoele, 161
Secondary "cytes," 14
Secondary oöcyte, 9
Secondary spermatocytes, 7, 8
Seessel's pocket, 199
Segmental arteries, 205, 206
Segmentation cavity, 90
Semicircular canals, 139
Seminiferous tubules, 7, 8
Sense organs, 198
Sense plate, 110, 111
Sensory axon, 135
Sensory neurons, 198
Seroamniotic cavity, 182
Seroamniotic connection, 224
Seroamniotic folds, 193
Seroamniotic raphe', 182
Serosa, 210, 224, 225
Sex cell cord, 154
Sex cell ridge, 154
Shell membrane, 170
Significance, of cleavage, 36
Significance, of meiosis, 19
Sinus rhomboidalis, 176, 178, 223
Sinus terminalis, 178, 191
Sinus venosus, 146, 147, 150, 152, 190, 191, 203, 207
Skeletogenous sheath, 82
Skull, 208
Small intestine, 118, 142, 203

Somatic mesoderm, 53, 58, 59, 60, 61, 62, 81, 96, 97, 121, 142, 177, 179, 182, 193, 222, 226
Somatic mesoderm, of chorion, 225
Somatopleure, 54, 60, 62, 81, 82, 121, 177, 179, 182, 192, 195, 224
Somite, 58, 60, 61, 79, 93, 97, 120, 142, 143, 144, 145, 177, 178, 179, 180, 181, 182, 194, 221, 222, 223
Somite, development of, 82
Somite, differentiation, 202
Sperm, 3, 6, 7, 8, 24, 105
Sperm, amphibian, 11
Sperm, aves, 11
Sperm, duct, 155, 214
Sperm, formation, 6
Sperm, head, 11, 12, 25
Sperm, in lumen, 8
Sperm, main piece, 12
Sperm, mammal, 11
Sperm, morphology, 10
Sperm, neck, 11
Sperm, penetration, 21
Sperm, perch, 11
Sperm, tail, 11, 12, 21, 25
Sperm, turtle, 11
Spermatids, 7, 8, 15
Spermatogenesis, 6, 7, 8
Spermatogonia, 7, 155
Spinal cord, 60, 99, 116, 131, 133, 178, 184, 197, 198, 223
Spinal cord, development of, 132
Spinal ganglia, 114, 228
Spinal nerves, 133, 198
Spinal nerves, distribution, 134
Spinal nerves, origin, 134
Spiracle, 126
Splanchnic mesoderm, 53, 58, 59, 60, 61, 81, 96, 97, 121, 142, 147, 177, 188, 193, 210, 222, 224, 225, 226
Splanchnocoele, 81
Splanchnocramium, 209
Splanchnopleure, 54, 60, 62, 81, 82, 121, 177, 179, 192, 194, 224, 226
Spleen, 203
Staining, vital, 43
Stomach, 118, 137, 141, 142, 154, 199, 203, 210
Stomadeal invagination, 111, 112
Stomadeum, 100, 118, 126, 140, 185
Striated muscle, 202

Subcardinal sinus, 150, 151
Subcardinal sinus, development of, 150
Subcardinal vein, 150, 151
Subcaudal fold, 98, 165, 186, 200, 223, 224
Subcaudal pocket, 165, 177
Subcephalic fold, 98, 163, 164, 165, 177, 178, 179, 186, 223, 224
Subclavian artery, 205
Subclavian, development of, 206
Subintestinal arteries, 101
Submucosa, 209, 210
Superior colliculi, 197
Superior vena cava, 150
Supraintestinal arteries, 101
Symmetry, planes of, 70
Sympathetic ganglion, 132, 136, 137
Sympathetic preganglionic neuron, 137
Sympathetic system, 136
Synapsis, 16
Synaptene, 15, 16
Systemic aortic arch, 154, 205
Systemic arch, 148, 149, 206

Tail, 195
Tail bud, 97, 114, 166, 180
Tail of the sperm, 11
Telencephalon, 129, 180, 183, 184, 194, 196, 228
Telencoele, 129, 131, 184, 196
Teleost, blastula, formation of, 88
Teleost, circulatory pattern, 101, 102
Teleost, cleavage, 88, 89
Teleost, egg morphology of, 86
Teleost, embryology of, 85
Teleost, eye, 99
Teleost, fate map of the early gastrula, 90
Teleost, fertilization of, 86
Teleost, formation, of the body shape, 97
Teleost, formation, of primitive gut, 95
Teleost, gastrulation, 90, 91
Teleost, mesodermal differentiation, 96
Teleost, neurulation, 94, 95, 99
Telophase, 14, 15, 18
Testes, 6, 7, 155, 214, 215
Tetrad, 16, 18
Thalamus, 196
Theca folliculi, 10
Thoracic air sac, 209

Thoracolumbar system, 136
Thymus glands, 140, 141
Thyroid gland, 141, 142, 186, 199, 202, 228
Thyrotropic hormone, 202
Thyroxin, 202
"Tids," 14
Torsion, 165, 180, 194, 196, 201
Trachea, 141, 186, 199, 202, 203, 209
Translocation, 16, 17, 18
Tricuspid valve, 204
Trigeminal ganglion, 180, 185
Trigeminal nerve, 136
Trochlear nerve, 136
Trophoblast, 218, 219, 220, 223
Truncus arteriosus, 146, 147, 149, 194
Tuberculum posterius, 115, 116, 184
Turtle sperm, 11
Tympanic membrane, 139
Types of vertebrate eggs, 14

Ulna, 207, 208
Ultimobranchial bodies, 140, 141
Umbilical arteries, 205
Umbilical cord, 228
Uterine canal, 227
Uterine endometrium, 227
Uterus, 226, 227
Ureter, 213
Urogenital system, 214
Utricle, 139

Vaginal canal, 227
Vaginal opening, 227
Vagus, 136
Vascular system, development of, 147
Vas deferens, 7, 155, 214
Vas efferens, 155, 214
Vegetal cells, 107
Vegetal hemisphere, 74, 104
Vena cava, posterior, 206

Ventral abdominal vein, 154
Ventral aortae, 146, 147, 152, 188, 190, 204
Ventral aortic root, 179
Ventral fin, 123
Ventral horns, 197
Ventral lip, 45, 77, 93
Ventral lip, of blastopore, 46
Ventral mesentery, 188
Ventricle, 146, 147, 167, 189, 190, 194, 195, 203, 204, 207, 211
Ventricle, brain, 130
Vertebrae, 202
Vertebrae, development of, 143, 144
Visceral arch, 119, 186, 201
Visceral arch, formation of, 119
Visceral cleft, 119, 182, 186, 194
Visceral grooves, formation of, 119
Visceral pericardium, 188
Visceral peritoneum, 210
Visceral skeleton, 209
Vital staining, 43
Vitelline artery, 191
Vitelline membrane, 25, 69, 87, 170
Vitelline vein, 146, 152, 182, 191
Vitreous humor, 138
Viviparous, 23
Vocal cords, 202

White matter, 132, 133, 198
Wings, 196, 208
Wolffian duct, 155, 203

Yolk, 13, 37, 60, 87, 89, 93, 95, 96, 97, 102, 121, 165, 170, 193
Yolk-laden endodermal cells, 113
Yolk plug, 46, 109
Yolk-plug stage, 109
Yolk sac, 23, 167, 192, 193, 207, 224, 225
Yolk sac, umbilicus, 192

Zona pellucida, 216, 218
Zygote, 24
Zygotene, 16